MW00974282

# Construction Operations Manual of Policies and Procedures

# Construction Operations Manual of Policies and Procedures

**SECOND EDITION**

**Andrew M. Civitello, Jr.**

McGraw-Hill, Inc.
New York San Francisco Washington, D.C. Auckland Bogotá
Caracas Lisbon London Madrid Mexico City Milan
Montreal New Delhi San Juan Singapore
Sydney Tokyo Toronto

*This manual is dedicated to the men and women who have built our country, and who will bring the construction industry into the next century.*

**Library of Congress Cataloging-in-Publication Data**

Civitello, Andrew M., date.
Construction operations manual of policies and procedures / Andrew M. Civitello, Jr. — 2nd ed.
p. cm.
ISBN 0-07-011048-4
1. Contractors' operations. I. Title.
TA210.C58 1994
624'.068—dc20 93-2651
CIP

First edition published 1983 by Prentice-Hall.



1 2 3 4 5 6 7 8 9 0 KP/KP 9 0 9 8 7 6 5 4

P/N 011235-5
PART OF
ISBN 0-07-011048-4

*The sponsoring editor for this book was Larry Hager, the editing supervisor was Peggy Lamb, and the production supervisor was Pamela A. Pelton. It was set in Century Schoolbook by McGraw-Hill's Professional Book composition unit.*

*Printed and bound by The Kingsport Press.*

# Contents

# Preface

As a professional manager, you realize that uniform techniques for planning, operating, and controlling that all levels of management can understand and *apply* are absolutely essential for risk management, profitable operations, and growth. A uniform method that is understood by all is a prerequisite for proper management. You also know that nonplanning encourages lack of uniformity, inconsistency, and incomplete, rushed responses to daily operating problems.

By clearly documenting your objectives and procedures, you take the most effective first step in the systematic and efficient instruction of all your managers, supervisors, and employees at all levels. Beginning with yourself, you will help each person by showing them how to be most effective at their jobs. You will help the organization for the same reason.

Construction is a game of incredible risks. Any problem at any construction site can quickly turn into one worth tens-of-thousands of dollars if it is not handled quickly, decisively, and correctly. A major problem can either be contained or be allowed to explode through mishandling or even personality conflicts. In the typical project and company management structure, too many people have vast amounts of "informal" authority—that dubious ability to commit the organization (through actions or *inactions*) that may have little or nothing to do with that individual's title or formal responsibilities. Most organizations simply rely on a superintendent's or project manager's personal skills—however complete or incomplete that person's training might be.

But we in construction go on starting project after project too often without applying the most positive influences that we can. Effective management successes last as long as one person's memory—if coincidence has allowed that person to use the experience again (rather than another person on another assignment).

We all accordingly realize that it is "obvious" that the trial-and-error method should not be relied upon as the primary method of instructing our employees, but for some reason, it is just that approach that is much too often being allowed to perpetuate. If you work in one of those rare companies that actually have some kind of manual or formalized instruction effort, the odds are that much of the information has fallen out of date or has otherwise fallen into disuse. It may be difficult to understand, difficult to apply, and difficult to monitor for its effectiveness.

So you've actually known for some time that you need a complete Operations Manual. But even if you've decided to make the necessary commitment to spend the thousands of hours required to write and produce the product, where and how do you begin? Even if all the policies and procedures are themselves effective, how do you know that they are being properly communicated?

Start by saving yourself and your company all that time and effort by implementing the *Construction Operations Manual of Policies and Procedures*.

# Acknowledgments

My deepest appreciation goes to:

- My father; Andrew Civitello, Sr., for the obvious, and
- My uncle; Edgar Civitello, for throwing me to the wolves as early as he could, and who, besides my father, was my most influential mentor.

Together, they are by far the most responsible for my professional development (or its really all their fault, depending upon how you look at it).

# Introduction

Most books written on construction methods, procedures, management techniques, and negotiating strategies rely heavily on theoretical detail. They attempt to discover where profit and losses occur and to suggest some possible means of improvement.

This manual begins where those "suggestions" end. It is a decisive action plan that directly turns attitude into action. It is not a book on how to make your operations manual—it *is* your operations manual. It is the result of the development of the most effective methods and operating specifics that have been proven over again by some of the most successful construction organizations throughout the country.

With ever-increasing specifications, heaping mounds of contract boilerplate, proliferation of computers, and the barrage of government regulation, it is crucial to report all categories of field information accurately, quickly, and completely. The success of failure of this effort translates directly into either containment of costs or into huge financial losses. Dramatic increases in the cost of doing business dictate that operations must be streamlined for maximum efficiency. Duplication of effort must be eliminated if overhead is to be controlled. Exponential increases in the incidence of arbitration and litigation mean that if you are to reach the top and *stay* on top, *all* your records must be timely, accurate, and complete.

If your company already has an operations manual, take a second look. Don't risk falling into the trap of letting your current methods becoming obsolete. Realize that the company's self-improvement program must be an on-going process, and that we must all constantly work to improve, correct, modernize, and refine *all* areas of operations in order to maintain levels of sophistication above that of your competition.

Large companies will have specialized staffs with limited ranges of responsibilities; those who purchase only purchase, those who estimate only estimate, and so on. Small to mid-sized companies necessarily combine many types and ranges of responsibilities in the same individuals. Those who estimate or bid a project, for example, may go on to become the superintendent, project engineer, and/or project manager.

What all this means is that the individual functions to be performed are very consistent company-to-company, but their combinations of assignments to individuals within those companies varies greatly.

The good news, then, is that the *issues* and operating ideals affecting construction are realy very similar for nearly all of us. The construction contract on a $1 million project, for example, will be nearly identical to that for a $20 million one. The plans and specs will differ in their technical and design content, but the issues, language, procedures, and decision theories are virtually the same. Every project has a project manager, superintendent, project engineer, and administrative support—whether or not the particular individuals realize that is what they are actually doing. The specific responsibilities being performed by an individual may have more to do with that individual's talent, predisposition, and experience than with any formalized job description.

The Manual's approach, then, is to segregate the activities of the construction team by primary function. Each of the resulting sections is treated in specific detail, and its relationships with the other functions demonstrated.

However the organization is formally arranged, those individuals responsible for specific tasks can use the relevant section of the Manual necessary for the moment—whether it happens to be the same function repeated every day by a specialist, or if it is one of several functions to be performed either every day or intermittently by an individual with a wide range of responsibilities. In all cases, the respective function will be performed consistently throughout the organization, and the individual performing it can clearly see its fit within the rest of the procedural system.

*This* manual is packed with over 300 pages of those procedures, methods, forms, letters, strategies, and tactics that continue to win creatively. They are expressed in a style of writing that is concise, easy to read, and well within the grasp of busy executives, managers, and employees. The material explains in detail the step-by-step "how" of directing and operating construction for profit. It is sharply focused on the facts; eliminating extraneous unnecessary detail.

The *detailed procedures, full-size forms, word-for-word letters, checklists, and form letters* provided are exhaustive. They are organized and presented in the best manner that will allow all company personnel to set and get what you want consistently, and to efficiently focus your time, energy, and creativity.

# What This Manual Will Do for You

If you have public or private construction contracts, this manual will help make your operation more profitable. Whether or not you are computer oriented, you'll be given economical, effective systems for planning, operating, and controlling. Weak areas in your operation will be flagged, and immediate action is explicitly described in a step-by-step, how-to program. Each employee will be given a quick reference to all company procedures, and each will be given the ammunition needed to do his or her job better.

*Construction Operations Manual of Policies and Procedures is a:*

- Complete Office Manual
- Project Engineering Manual
- Field Superintendence Manual
- Project Safety and Loss Control Manual
- Contract Risk-Control Manual
- Comprehensive Project Management *System*

In each of these areas it provides:

- *Ready-made, full-size forms*, with complete instruction.
- *Step-by-step direction* to handle every operating situation.
- *Complete Checklists* that consolidate every important component of a given issue.
- *Word-for-word letters* that have proven themselves to be the most effective in handling every situation. Each is loaded with obvious and subtle language designed to control the short- and long-term of each issue.
- *Form letters* that simplify and apply efficient handling of the many routine processes.

The Manual has been produced from the ground up with the time-squeezed professional in mind. It is organized into practical operating sections that allow each individual to focus on only those areas that are in that person's direct responsibility. The superintendent in a large organization, for example, need only to refer to the Section "Site Superintendence" for guidance on the majority of his or her responsibilities. Whenever desired, however, that person can easily review any other section (Project Engineering, for example) to see how his or her responsibilities coordinate with and supplement the field operations. A project manager in a smaller organization can put his or her "Project Engineer," "Contract Administrator," and "Superintendent" hats on all in the same day by referring to the appropriate sections of the Manual.

*Construction Operations Manual of Policies and Procedures* will help you reduce errors, eliminate duplication, and give you the power to take control of your contracts and subcontracts in a way that will improve cooperation on the parts of subcontractors, owners, and design professionals. The first day that it is implemented, your construction projects will begin to run more smoothly. You will be relieved of all the frustrating procedural details and attendant constant decision making. Your contracting system will almost automatically anticipate problems, apply quick, complete solutions, and protect yourself and your company against the otherwise disproportionate risks that are waiting around every decision.

## A Special Feature

All *Word-for-Word Letters* and *Form Letters* have been consolidated on 5 1/4" and 3" floppy diskettes for immediate use with your specific word processing software.

Although the hard copies in the Manual can each be photocopied or otherwise reproduced as they are, their inclusion on these diskettes fits them all immediately into your existing software system. Simply use the files as any other word processing document.

# Brief Contents



## Special Features:

- (1) 5 1/4" and (1) 3" floppy diskette; each with copies of all:
  —Form Letters
  —Word-for-Word Letters
- Prepared for immediate use with your specialized word processing software.

## Checklists



## Forms

## Form Letters

## Word-for-Word Letters

# An Important Note

The letters, forms, checklists, and instruction throughout this Manual are important examples of the way in which their respective situations have been handled effectively by construction professionals. Many operations and responses can be handled precisely as indicated. It is important, however, to be aware that laws vary across state lines and that trade practices may differ depending upon the geographic location of your business and/or jobsites. Beyond that, all operations boil down to your own business decisions regarding appropriate responses to complex situations. The more important notifications, reservations of rights, direct statements, and so on are critical business communications that must be considered carefully in each application. Before dramatic action is taken in serious problem situations, for example, the advice of a competent attorney should be sought.

*Construction Operations Manual of Policies and Procedures* has been designed for fast reference by busy professionals. If for any reason you wish for additional clarification or have specific concerns not directly addressed here, we will be happy to consider your written questions addressed to:

Operations Manual
PO Box 190
Bethany, CT 06524

Your comments, remarks, and suggestions regarding the *Operations Manual* and its presentation are most welcome.

Section

# 1

# Company Organization and Quality Assurance Program

## 1.1 Section Description

Section 1 is an introduction to the company. It includes the complete description of the company's organization, a statement of who we are and what we are doing, and the reasons for doing it. It summarizes authorities and chains of command and describes how formal and informal communications will be conducted throughout the organization.

## 1.2 Company Purpose: Operating Statement

The purpose of the company is to provide construction and construction management services of the highest professional standard. It will be accomplished in a manner that is consistent with the intent of industry standard agreements and contracts, which may be modified for specific projects, owners, and design professionals.

The company will operate with the objective of generating profit consistently above the industry average, but it will operate in a way that *always adds value* to the process.

Every executive, manager, supervisor, administrator, and employee should guide his or her day-to-day operating decisions with specific consideration of performing any task or solving every problem in a way that:

- Is expedient
- Finishes the item completely—the first time
- Enhances relationships with all parties involved
- Adds to the reputation of the company and to personal stature

While at times, for given situations, these objectives may seem mutually exclusive, striving for their attainment in *every* situation will consistently result in efficient work, creative solutions, respect, and personal satisfaction.

## 1.3 Organization Structure

### 1.3.1 Corporate Office

The company office location and contacts are as listed in the Construction Operations Manual Transmittal provided in the introductory section of the company manual.

The functions of the corporate staff include:

- Determining objectives and policy
- Performing the primary disciplines of Business Development, Finance, and Accounting
- Coordinating and integrating all production
- Providing guidance and assistance to all company members
- Resolving situations that are beyond the ability or authority of the individual project staff members

### 1.3.2 Project Field Offices and Staffing

Project staff functions include:

- Project Management
- Site Superintendence
- Project Engineering
- Project Accounting
- Administrative and Secretarial Services

The responsibilities for the functions of these categories may be assigned to individuals, and even to staffs for large projects, or may be combined in fewer individuals for smaller or less complex projects. The project manager on a smaller project, for example, may also be responsible for performing the functions of the project engineer and may even be responsible for varying degrees of procurement and estimating. The details of these relationships begin in this section, but they are developed throughout the Manual.

The project offices are located at the respective jobsites. For small to mid-sized projects, the on-site staff will be limited to the general superintendent, and possibly the area superintendents and/or trade foremen as appropriate for the specific project. For projects that are large and/or complex, appropriate degrees of project administrative personnel will also be located at the site. These staff positions may include the project manager, project engineer, office engineers, project accountant, and project secretary.

The specific organizations of key functions and personnel, along with their relationships with the other corporate operations, are covered later in the section.

### 1.3.3 Operational Objectives

The specific organizational structures in the following exhibits have been designed with objectives of effectiveness, efficiency, and elimination of redundancy.

Comprehensive treatment of the complex issues faced each day in the industry can lead to complicated, detailed activities that can span many operational levels. It is important for each individual throughout the organization to appreciate that all company procedures have been established, not for their own sake, but have been instituted to achieve specific objectives. How an individual's function fits within the overall objective may not always be clear.

The ideas of "teamwork" and "cooperation" cannot be treated as the clichés they can appear to be. They must consistently be applied in the truest sense if the company objectives are to be met, and if those individuals meeting those objectives are to develop personally and professionally.

The larger the organization, the more difficult it may be for the individual to maintain a sense of individual accomplishment. Direct lines of cause and effect may become clouded. If any procedure at its institution or during its development over time begins to take on the appearance of a bureaucracy for its own sake, it is *crucial* that those involved in its process continually evaluate its effectiveness in *achieving objectives*. Appropriate corrections to keep activities streamlined toward *goal achievement* should become habitual in the awareness of each individual. Efforts toward this end should be openly rewarded whenever an opportunity arises.

## 1.4 Organization Charts

### 1.4.1 General Description

The organizational structures that follow display the individual components of the respective operation. In large companies, each function will have individuals or even staffs assigned to it. Smaller organizations will necessarily have fewer individuals involved in its administration, but the *functions* performed should be the same. The functions as listed will accordingly be combined in the responsibilities of those individuals. The formulas determining those combinations will be most often based on the specific person's capabilities, inclinations, and initiatives. A project manager (PM) in one company, for example, may also do project engineering, or in another company, may do estimating or purchasing. In still another company, the PM may also serve as the site superintendent.

Observe the functional relationships, and visualize the appropriate individual responsibility assignment within your own company.

Specific jobsite structures will be treated in Sections 1.5 and 1.6.

### 1.4.2 Corporate Organization

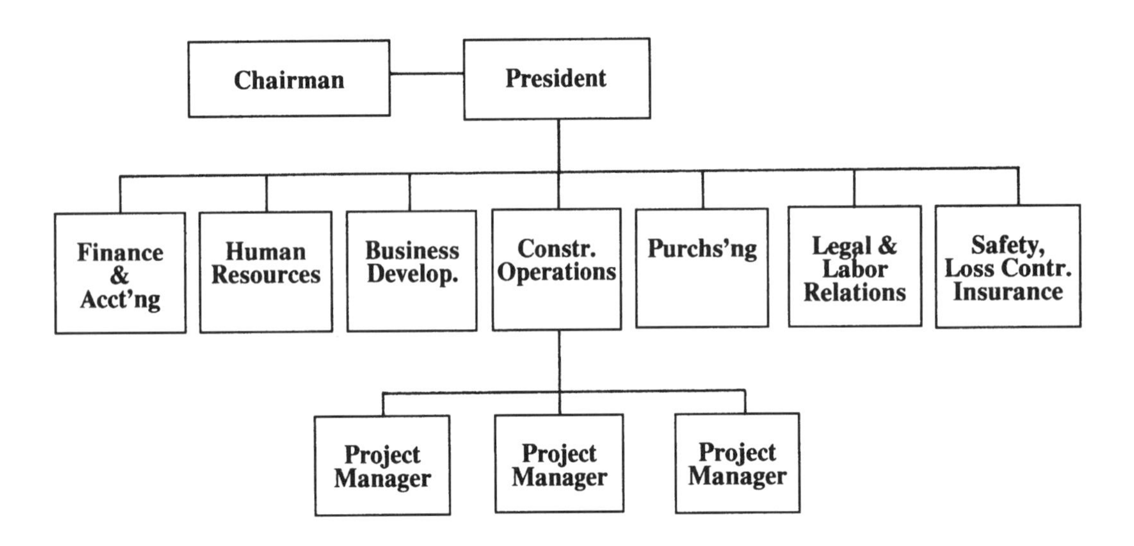


## 1.5 Jobsite Structures—Large and Small Projects

### 1.5.1 General Description

The detailed descriptions in this section assume a large-project structure for ease in clarifying the individual relationships of the particular line and staff operations. A large and/or complex project may have an individual and even staff assigned to each function—depending strictly upon the needs and responsibilities of the respective assignment.

Smaller projects will require fewer individuals, but the complete list of functions performed will be the same. Examples of ways in which duties may be combined include:

1. Project Manager/Site Superintendent/Project Engineer/Scheduler in a single individual
2. Project Manager/Purchasing Agent working with a Site Superintendent/ Project Engineer, each working with centralized scheduling
3. Project Manager/Project Accountant working with Site Superintendent/ Project Engineer/Scheduler

The specific mix will be determined by the requirements of the project and the dispositions and capabilities of available individuals.

The descriptions that follow are summaries of the overall fit of the respective functions within the project structure. Details of the activities themselves are treated throughout the Manual.

## 1.5.2 Typical Project Organization Chart—Large Projects

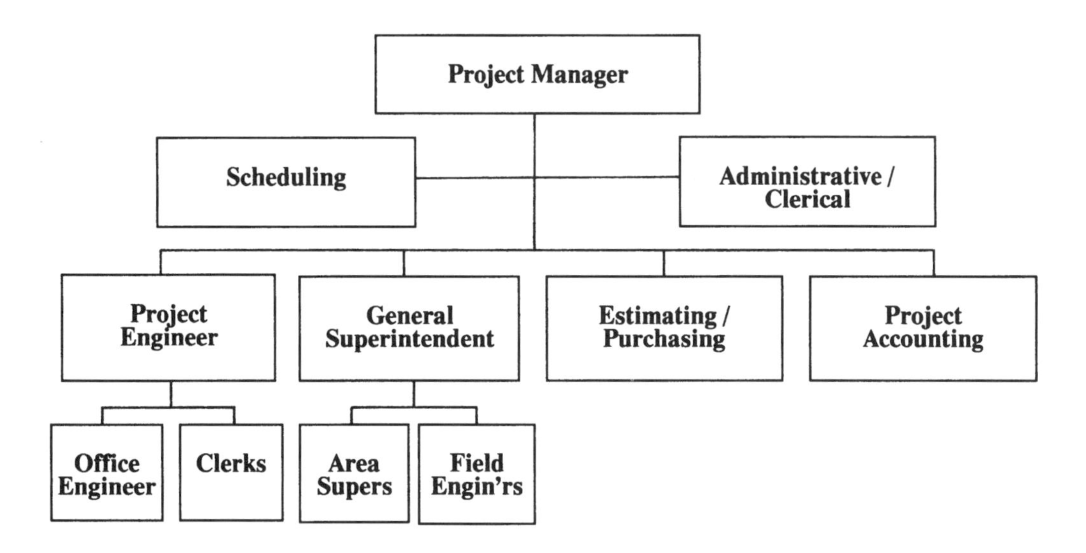


## 1.5.3 Concept, Organization, and Attitude

Construction Operations is the mechanism by which all company objectives as described in its operating statement will be achieved. It is what we sell in our Business Development efforts, and is the force that generates *all* revenues. Individual project success as defined by cost/profit, time of completion, and quality thus translates directly to success of the company and of the individual. Accordingly, *the purpose of virtually every position throughout the organization is to support construction operations.*

The summary descriptions (1.5.4 through 1.5.9) of the individual and/or staff responsibilities of the various project functions display their relationships. Their specific activities and responsibilities are treated in great detail throughout the Manual.

Efficient operation dictates the need for line and staff communications, decision-responsibilities, and authorities. It is important to acknowledge that the *graphics* of organization charts tends to accent the *impression* of "superior/subordinate" relationships throughout. The truth is, each line responsibility and staff function takes its turn repeatedly in becoming *the* critical operation of the moment. The Project Manager is considered to be the focal point of coordination for the information generated by all functions and is the point-person who communicates that information to the corporate office and to the outside world. Through the Project Manager, *all individuals share equal importance and relevance to the success of the project effort.*

That statement is not a cliché about the successful project management effort. It is a profound realization that each component of the entire construction process is crucial to the success of the whole, and each can carry with it extreme liabilities for the entire project staff and the company itself if that function is not performed as well as it can be—on time every time.

Those who insist on operating with the idea that they are somehow individually more important to the effort than the others will reach the limit of their effectiveness very quickly. On the other hand, those who continually help others to do *their* jobs better will grow quickly both through their own personal and professional development and through the support of their peers born out of appreciation and respect.

### 1.5.4 The Project Manager

The Project Manager (PM) is the individual charged with responsibility for the complete and satisfactory execution of the entire project. In very large organizations divided into regional areas, the PM may report to an executive or other person who would in turn report to the Operations VP. In most companies, however, the PM reports directly to the Operations VP or to the company President. In many small companies, the PM may *be* the Operations VP or the President.

In any case, the Project Manger is the company representative who deals with the outside world (owner, design professionals, vendors). His or her actions, as such, will be regarded as the company's. However good or bad that person appears to be the company itself appears to be.

General duties of the Project Manager include:

- Assisting in development of the project staff
- Coordinating on-site with corporate office activities
- Organizing and overseeing jobsite administration
- Organizing and coordinating field supervision
- Assisting in the procurement of subcontractors and suppliers
- Developing, monitoring, updating, and communicating the progress schedule and its periodic revisions
- Managing the direct labor force and maintaining labor relations
- Managing subcontractor schedules, quality-of-work, coordination with other trades, and payments
- Coordinating cost-progress targets with production

- Creating and maintaining a safe/secure jobsite environment
- Identifying and resolving all changes
- Establishing and maintaining relationships with the Owner, design professionals, building officials, local businesses, and police and fire departments

### 1.5.5 The Superintendent

The Superintendent is responsible for the supervision of all field activities related to the physical construction. He or she reports to, and carries out the direction of, the Project Manager with respect to field operations, and thereby directs the daily progress of the work. The jobsite work force is made up of an army of specialists. The Superintendent must assure that the combination of their components results in a cohesive product that achieves the required quality and is completed in the shortest possible time. To that end, the Superintendent must continually work to assure adequate staffing of the work force, sufficient supply of materials, and complete information as necessary to assemble the product, and must plan for all these things sufficiently enough in advance so as not to interfere with the progress of any one component. General duties of the Superintendent include:

- Generating, securing, or otherwise confirming all information needed to create, monitor, and modify the progress schedule on a continuing basis
- Developing the progress schedule with the Project Manager
- Participating in scope reviews of the various bid packages to properly coordinate their respective interfaces and ensure that nothing is either left out or bought twice
- Working with the Project Manager to develop and administer the site utilization program, site services, security arrangements, and other facilities and arrangements necessary for appropriate service to the construction effort
- Identifying field-construction and work-sequence considerations when finalizing bid package purchases
- Monitoring actual versus required performance by all parties
- Determining whether subcontractors are providing sufficient work force and hours of work to achieve performance commitments
- Monitoring the performance of the company's Purchasing and Project Engineering functions to ensure that all subcontracts, material purchases, submittals, deliveries, clarifications, and changes are processed in time to guarantee jobsite arrival by, or before, the times needed
- Directing any company field staff
- Being thoroughly familiar with the requirements of the general contract, thereby identifying changes, conflicts, etc., that are beyond the scope of responsibility
- Preparing daily reports, job diaries, narratives, and all other regular and special documentation as determined by the company and by the project needs

### 1.5.6 The Project Engineer

As the Superintendent manages the physical construction of the project in the field, the Project Engineer in a sense builds it first—on paper. The Project

Engineer is responsible for coordinating the complete construction and administrative requirements of the various bid packages; organizing the complexities into a management system that will establish, monitor, and follow-up on each vendor's compliance; and orchestrating the information flows needed for each component and system to be incorporated into the project—all within the time frames needed by the progress schedule. Strict attention to detail, a profound appreciation for time constraints, and a sober understanding of the dividends reaped through proper documentation all add up to an effective Project Engineer. Duties include:

- Working with the Project Manager, Superintendent, and the estimating/purchasing effort in the final development of the various bid packages and the baseline construction schedule
- Working with the Project Manager and Superintendent to develop site utilization programs
- Establishing, maintaining, conducting, and policing detailed procedures for the submittal, review, coordination, approval, and distribution of shop drawings, samples, etc.
- Establishing and maintaining all project engineering files relating to subcontract and bid package records, plans, specifications, changes, clarifications, and as-built documents
- Reviewing all vendor schedules of values and preparing the general schedule of values as coordinated with the Project Manager
- Managing the periodic requisition procedure for review, submittal, and payment
- Expediting vendor estimates and proposals and preparing appropriate company estimates and proposals for changes to be submitted to the Owner
- Determining appropriateness and preparing subcontractor change orders to be processed through the Project Manager
- Evaluating subcontractor payment requisitions relative to actual work performed
- Working with the Project Manger, Superintendent, and Project Accountant to prepare the general requisition and follow it through to payment

### 1.5.7 The Project Accountant

The function of the Project Accountant can vary from strict application of technical accounting to filling a variety of roles important to the success of project administration. On large projects an individual assumes the role, with the Project Manager or Project Engineer assuming it on smaller projects.

The Project Accountant is the financial watchdog over project administration. He or she will be involved with monitoring cost performances, verifying cost information from all internal and external sources, maintaining all appropriate records, and reporting both to accommodate the technical requirements of the central accounting effort and to quantify performance. Because of its nature, the position naturally lends itself to that of Office Manager. In this role, the Project Accountant's responsibility can expand to that of auditor of company procedure. Responsibilities include:

- Establishing all vendor files relating to contracts, billings, payroll and labor records, and material costs

- Maintaining job-cost information on a current basis and producing cost-progress reports, highlighting areas potentially exceeding estimates in time to implement corrective action
- Enforcing compliance with payroll and EEO reporting and securing lien waivers from vendors of all tiers as condition of payment
- Performing periodic audits of office procedure to monitor company performance
- Reporting project financial information to central accounting

### 1.5.8 Planning and Scheduling

Too many people equate the scheduling function to that which simply produces the finished reports. Actually, planning, scheduling, monitoring, and updating constitute *the* fundamental activity through which all others follow. Sequences affect estimates, schedules become contract and subcontract commitments in their purchasing and performance phases, extended delays dramatically affect multiple trades, and the time of completion affects everyone's wallet.

Chronically missed schedules cause extremely disproportionate costs and interferences resulting in economic loss, problems that permeate multiple organizations, and bad reputations that are difficult to shake off. In contrast, those companies that actually give the scheduling function the effort it deserves enjoy higher profits, better relationships, and improved reputations as direct results.

The first purpose of a comprehensive construction program is to establish the quickest, most effective methods of assembling all construction components. That requires the vision, expertise, and determination of the Estimator, Project Manager, Project Engineer, Superintendent, and the force of specialty contractors if all desired objectives have any hope of being met with a minimum of conflict. The Scheduler commits the plan to paper and keeps the information distributed to everyone affected, actually or potentially.

From that standpoint, planning and scheduling remains *everyone's* responsibility. The time-status of every component must habitually become the focal point around which all other information is arranged. The potential effect of *every* issue on the progress schedule must always be a key consideration throughout each issue's resolution.

Duties include:

- Identifying each major construction activity, its relationships with other activities, and all necessary support
- Correlating the activity list with the contract documents and the schedule of values
- Soliciting and confirming all information from the best combination of sources, incorporating it into the plan, and distributing it in a timely manner
- Monitoring actual progress relative to planned progress continually, assessing its actual and potential impacts, displaying cause and effect relationships, and determining necessary corrections
- Monitoring the plan's implementation and maintaining all documentation relative to good and bad performances of all parties in a manner that is complete,

correct, and well correlated and can be used most effectively by the project management, finance, and legal efforts

### 1.5.9 Project Management Services, Staff, and Support

The recurring theme throughout the Manual is that the success of the production effort *is* the success of the company. To accommodate that reality, all other company operations exist to support production efforts. The individual line-and-staff operations themselves are treated exhaustively throughout the Manual. Here, it is sufficient to emphasize that all processes are equally important, and that none exist for their own sakes.

Estimating, for example, needs to be aware of realistic, achievable production rates. Purchasing needs the ability to verify whether promised performances are achievable, and if so, by what specific means that sub-vendors are willing to commit to contractually. Subcontracts are to be secured in time to guarantee smooth flow of the work. All information relating to project engineering, project accounting, planning and scheduling, and even office administration must be timely, accurate, and complete. All this must be accomplished within the spirit of efficiency and managed in an environment of cooperation.

The prevailing attitude throughout the company should be one that encourages—even rewards—early warning of potential problems and conflicts. There's no better guarantee that small problems will become large ones if they are not dealt with early, or when staff members fear a "kill the messenger" disposition or penalty in any form for bringing bad news to the attention of superiors.

With the complexities, pressures, and problems that are thrown in our faces each day, we must all constantly remind ourselves that we're all on the same side and that the clock never stops. The concept of a complete team approach must be taken seriously as it deserves to be. Those who ignore that concept will plateau quickly, and their careers will soon follow.

## 1.6 Document Generation, Signing Authorities, and Communication

### 1.6.1 Document Authority

Strict lines of formal authority can be detailed in every organization in a straightforward manner. In the construction industry, large networks of informal authority structures can develop if they are allowed to. Those in turn can be further complicated by incorrect *perceptions* of both formal and informal authority.

Every act, deed, and remark made by a wide range of individuals carries some potential to commit the organization. Examples range from the clear language in a contract executed by a company officer to a remark by any member of the project team made at a job meeting or while touring the site. Clearly defined procedures can be thwarted when the *actions* of the parties constructively alter the "real" relationships. If, for example, a Change Order is not formally executed but the work is completed—and paid for—it can dramatically alter the legal application of the entire Change Clause and even the Dispute Resolution Clause of the general contract.

All this simply means that in order to contain the significant risk of unintentionally or otherwise inappropriately committing the company or creating additional liability, lines of authority as determined in the formal project team structure must be strictly observed and respected. Each type of communication with the outside world, ranging from the most routine communication to the most sensitive issues, must be handled by those with the confirmed authority to do so, and in the way prescribed by company policy. Situations handled by one individual must not be compromised or undone by uncoordinated efforts of others.

Finally, every type of communication has had a policy determination for the way in which that issue should be handled. Those procedures have been designed with the benefit of experience under-fire over a long period of time. It is the intent of the policy that the manner in which all communications are conducted is as directed and that the intent of the procedure is preserved. In those areas where form language and/or the form itself is provided, it should be used. When the appropriateness of a stated policy approach is of concern for a specific situation, final treatment of the issue should be reviewed with the immediate supervisor.

## 1.6.2 Document Types and Signing Authorities

A list of various routine documents between the company and the outside world typically generated repeatedly throughout each project follows. Beside each document are those respective personnel authorized to generate, modify, or sign (internal document procedures are detailed elsewhere throughout the Manual).

**Contracts, Subcontracts, and Contract Modifications**

| | |
|---|---|
| Owner/Company (GC, CM) | President |
| Company/Sub-Vendors | Corporate Officer |
| Change Orders—To Owners | Corporate Officer |
| Change Orders—To Sub-Vendors | Corporate Officer |

**Project Management, Engineering, Accounting**

| | |
|---|---|
| Schedule of Values Submission | Project Manager |
| Payment Applications (To Owner) | Corporate Officer |
| Overbudget Authorization | Corporate Officer |
| Sub-Vendor Schedule of Values Approval | PM/PE |
| Sub-Vendor Requisition Approval | PM/PE |
| Process Sub-Vendor Requisition | PA |
| Shop Drawing Reviews/Transmittals | PE |
| Schedule Transmittal/Notifications | PM/PE |

**Documentation and Correspondence**

| | |
|---|---|
| Job Meeting Minutes | PE |
| Routine Correspondence | PM/PE/PA |
| Backcharge Notices PM/PE/Super | |
| Delay Notice—To Owner | PM |
| Delay Notice—To Sub-Vendor | PM/PE/Super |
| Notification of Pending Change Order | PM/PE |

| | |
|---|---|
| Change Order Proposal | PM/PE |
| Notification of Work under Protest | PM |
| Work Stoppage | Corporate Officer |
| Notification of Claim | Corporate Officer |

## 1.7 Quality Concept and Quality Policy

### 1.7.1 Concept of Quality

It is often difficult to consider the subject of quality as applied to both the tangible field construction and management and administrative processes without at least a measure of cynicism. I'm certain that the reasons for this include everyone's experience with the failure to satisfy everyone, primarily because each individual involved happened for the moment to have a different idea of what "acceptable performance" was supposed to be. The words have been overused and abused, without so much as a clear definition of what everyone's been complaining about. Add to that the idea that "quality" somehow takes more time (and therefore more money), and work (both field and administrative) is even started with the specific intent of cutting the proverbial corner. We are in an incredibly time-squeezed business. There is never enough time to do the job right. There always seems to be, however, enough time to do it again—after we've added time to argue about it.

This is not just a speech. It is a real acknowledgment that the failure to accomplish tasks correctly, completely, and on time—and to do them just once—causes continual disruption, creates those incessant fires, and directly causes those arguments, delays, rework, and all the other problems that explode budgets, destroy schedules, set the company back, and put people's careers on hold.

The concept of "quality" is not the vague, subjective, somewhat intangible ideal that most people actually equate with the word. Words like "high quality," "smooth," or even "satisfactory" do more to confuse than to clarify.

To the contrary, the idea of "quality" is *very* specific. It is simply to meet or exceed the stated requirements, to do so the first time, and to do so in every situation. If stated requirements are unclear or otherwise inappropriate, they must officially be changed. The new, clear requirements can then be met completely.

If all the requirements of a luxury sedan are met, then it is a quality vehicle. If all the requirements of a basic pickup truck are met, it is a quality vehicle. It has nothing to do with luxury. Luxury, or its absence, is spelled out in the vehicle's requirements. If each conforms strictly to its requirements, both are of high quality.

The second misconception about quality is that it to varying degrees is intangible, and is therefore not (or not easily) measurable. In fact, the opposite is the case.

If the specification, for example, uses words like "flat," or "reasonable promptness," it is not a quality specification, and the requirements cannot be "met" without significant interpretation (and probable argument). Performance relative to the requirement cannot be properly evaluated. Those kinds of specifications must be changed to read "flat to within 1/8″in 10′ in any direction," and "within ten workdays." The requirements become tangible and specific, and relative performance can easily be *measured.*

Defining a punchlist completely—the first time—is quality administration; completing it expeditiously—once—is quality management.

The whole idea of quality is one of the most difficult to push through the clichés and misconceptions of years of misapplication. It is clear to me that some beliefs are so ingrained that they cannot be changed simply by suggesting that they're wrong. As simple as they are, the ideas in this small section are a dramatic departure from the customary point of view. They must, however, become no less than the basis for a change or improvement in the company culture—our fundamental approach to our entire sphere of work. We must all learn to see things in terms of those who must finally wind up doing the job and to explain quality in terms that cannot be misunderstood. Implementation must be people- and process-oriented, specifically defined, and measurable. If they are not, the definition of the task must be nailed down and the requirements quantified before it is begun. Fire fighting must be replaced with prevention of mistakes. Every activity must carry with it:

- A tangible, accurate description of the completed item and time of performance
- The ability to measure compliance/performance
- Delineation of the specific process and resources to be used
- Specific accountability for the outcome

We all have a tendency to gravitate to the site—to the place where the physical construction is being performed—to view the result of good or bad performances relative to quality. The presence or lack of that "luxury" component of quality will be visually and immediately apparent. It's an easy matter to locate the individual tradesperson who screwed the thing to the wall and demand that he, she, or their company fix it. It is a surgical strike at a symptom, and it too often requires disproportionate amounts of company resources to correct—even if it is not "our" problem. It's almost like stating that the worst crimes occur in the ghetto or are created by that mugger in Central Park. We quickly prosecute them because the individual is relatively easy to grab, once identified, and can offer little defense for the $200 crime. The problem is a pinpoint. The solution is immediately clear and requires the efforts of few people. Results are immediately measurable, and success is clearly defined. And so we focus all these energies on solving the relatively small issue while a white collar criminal is walking off with millions.

We managers walk blindly past our own accounting, business development, estimating, purchasing, and project management functions on the way to the project to look for the errors in the field. And we find loads of them. The truth is, the people on the site can, and do, contribute to lowering unnecessary costs relating to rework and other problems of substandard construction when they are brought out. The much larger problem, though, remains the fact that all planning and creation is done somewhere else.

It is a much more difficult task to pin the fundamental cause of a problem on the fact that the bid package itself, and the resulting subcontract, did not meet the requirements of a "quality" product in the first place, or that our own obligations relating to times of performance or holding up some other part of *our* end of the bargain were not met. These other places are where we must spend the greater part of our efforts to reduce related costs. It is a guaranteed bet that the causes of most of the expensive problems are at the other end of pencils, telephones, and computers—not on the end where the hammers, shovels, and paint brushes are.

### 1.7.2 Quality Policy

It is the policy of our company to provide the highest quality construction services. To this end, we will meet or exceed all stated requirements of every endeavor in which we are involved, with respect to materials, labor, and administrative or time components.

Each subjective, unquantifiable requirement must be made tangible. We must guarantee that *every* requirement is clearly and specifically defined in terms that can be measured and communicated in such a manner that those responsible for its execution will clearly understand. If any requirement cannot be met, we must cause the requirement to be officially changed.

## 1.8 Business Development Approach and Project Participation

"Official" business development efforts are directed from the corporate office. Nowhere else but in construction, however, are the activities of so many individuals throughout the organization so important to the sustained ability of the company to secure new contracts.

Manufacturing industries, for example, typically rely on the success of the end-product, and on the size of the advertising budget. In our business, the project is our product. While it, of course, will long stand as testimony to our ability to meet the requirements, the additional component to complete project success and owner satisfaction is the *way* we meet those requirements.

It is the job of centralized business development to provide the trappings of marketing—the brochures, information, and presentation—but it is a major component of *everyone's* job description to be that of Key Salesperson. Every key employee must be constantly aware that his or her actions daily affect company stature.

Most clients recognize that construction can be a difficult process riddled with inevitable problems and conflicts. The way in which we all organize and execute our work, the cooperative, optimistic attitude displayed in our approaches to problem resolution, and our effectiveness in keeping our objectives targeted at client satisfaction while preserving the other project objectives will stand like a neon sign confirming our professionalism. These elements must be second nature, *primary* considerations in the performance of every task, every debate with Owner representatives and design professionals, and the final position in the resolution of every conflict.

The successful project, satisfied client, and relationships maintained with design professionals are our best advertisements—our best sales efforts.

As our projects are completed in this manner, the primary roles of our "official" Business Development effort will be the demonstration that our project personnel are the company's greatest advantage that any new client can secure.

## 1.9 Marketing Services and Support

### 1.9.1 Marketing Ideals

The primary objective of the marketing effort is to increase and maintain favorable exposure to potential buyers of construction services in the public, private, and institutional sectors. We thereby multiply our opportunities to be considered for construction assignments throughout the industry.

The marketing ideals of the company are to:

- Maintain corporate identity and image
- Operate with and promote professionalism
- Display competence, leadership, and effectiveness
- Communicate integrity

A consciousness of these ideals should be made part of everything we accomplish, and the philosophies should manifest themselves in our actions.

This section describes the objectives and responsibilities of all non-Business Development professionals throughout the company. Summaries of marketing aids available, some suggestions on their use, and your responsibilities follow.

### 1.9.2 Literature and Supplies

Each Project Manager and Project Engineer is responsible for maintaining adequate quantities of company literature and supplies. You will be advised of all marketing materials currently available, be supplied with an initial package, and be advised of replenishment sources and contact people. Do not run out before replenishing.

### 1.9.3 The Corporate Brochure

The corporate brochure has been designed to create a favorable climate and feeling about the company in the minds of potential users of our construction services. It promotes professionalism, diversification, relevant experience, management depth, and integrity.

From time to time, select distributions of brochures will be made, but the most effective presentation is *personal contact.* Be sure that you have an adequate supply of brochures on hand and personally distribute them to appropriate people at every opportunity.

Logical recipients include:

- Current clients
- Facility planners
- Architects and engineers
- Business executives
- Building or Building Advisory Committee members
- Developers

### 1.9.4 Newsletters

The Newsletter is designed for periodic selective mailing distribution. Its purpose is to update and supplement the corporate brochure. It is a good reason to "follow up" on a prospective client, and it presents an opportunity to further a potential relationship.

You will be advised of specific Newsletter mailings, and you will be expected to be familiar with, and to use, the Newsletter in subsequent personal contacts.

### 1.9.5 Business Cards

A sample of your business card is shown below. As with any other marketing aid, it does no good in your pocket. Use it at every opportunity.

Place
your business card
here

### 1.9.6 Contacts

Develop "secondary" personal contacts as information sources of lead information on available and upcoming projects in concept, planning, design, and bid phases. Some sources of these contacts may include:

- Architects and engineers
- Subcontractors
- Material suppliers
- Core boring and other testing companies
- Planning and Zoning agendas
- Trade unions
- Government and industry sources
- Local and state development agencies
- Banks and lending institutions
- *Dodge Reports* and other project information services
- National publications and magazines
- Newspapers
- Trade/industry magazines and journals
- Business owners and executives

### 1.9.7 Mailing List

When the corporate brochure or any other marketing aid is personally distributed, or when you mail it yourself, please note the individual and/or company contacted, and furnish this information directly to the company Director of Business Development. This information is needed to construct and continually update the mailing list for all our marketing programs.

### 1.9.8 Some Dos and Don'ts

1. Speak positively about our company always, but don't overstate your capabilities.
2. Set aside a predetermined number of hours each month devoted exclusively to client contact. Activities can include phone calls, lunches, and visits.
3. Know your clients' industries. Know and use their buzz words. Get their trade journals.
4. Participate in industry trade seminars.
5. Join organizations—as many as possible.
6. Know people on development boards.
7. Invite clients to see current projects. Then ask questions that you lead.
8. Never neglect to submit a bid at the last minute without being certain that you're not putting the Architect on the spot. He or she may be depending on you.
9. Be on time for appointments.
10. Drop clients' names.
11. If you see a client's name or photo favorably publicized, cut it out and send it.

### 1.9.9 Suggested Proposal Index

The suggested Proposal Index scheduled below is presented in order to give the Project Manager and company executives a list of possible topics to address in a project proposal that might favorably impress a potential client to further consider the company for a particular construction assignment. The information is designed to promote competence and professionalism. The proposal can include:

- Transmittal letter
- Bid format
- Company qualifications
  - Experience on comparable projects
  - Corporate management and organization
  - Statement of financial capability and bonding capacity
- Area considerations
  - Labor market
- Project approach
  - Management and labor approach
  - Preconstruction plan and program
  - Jobsite operations plan
- Project organization and résumés
  - Project staffing and organization
  - Position descriptions
  - Résumés of key personnel

- Project schedule
  - Overall plan and program
  - Review of options and considerations
- Subcontractor policy and list
  - Subcontract policy
  - Subcontract approach
  - Subcontract listing
- Construction equipment and tools
- Insurance
  - Meet or exceed minimum coverages
- Company brochures, newsletters, and other promotional materials

Section

# 2

# Company and Project Administration

## 2.1 Section Description

Administration as applied in this section of the manual is the housekeeping of the organization. It establishes the basic procedures for use of the manual, staff organization, basic setups of various files and facilities, and overall logistics of communications. From here, we will have a place to put everything developed in other sections of the manual.

### 2.2 Use of the Operations Manual

#### 2.2.1 Overall Approach

The manual has been organized to make its information as usable as possible to the greatest number of construction professionals. Although the apparent number of differences in the way that companies operate may appear staggering, the truth lies in the fact that the actual functions performed are very similar.

Large companies will have specialized staffs with limited ranges of responsibilities—those who purchase only purchase, estimators only estimate, and so on. Small and mid-sized companies will necessarily combine any number of specialized activities in the same individual. Those who estimate and bid a project, for example, are likely to be heavily involved with purchasing. In some companies, those same individuals may then go on to become the project manager. People with primary superintendent responsibilities may also be called on to purchase a few or many items. All hiring may be centralized, or each manager may have the authority to hire his or her own staff.

All this simply means that within our own company and all those companies with which we deal, the specific *combinations* of job assignments may vary greatly, but the individual functions and assignments themselves are remarkably consistent. The *issues* and operating ideals affecting construction are really very similar for nearly all of us.

The construction contract on a $1 million project will be nearly identical to that on a $20 million one. The plans and technical specifications will be different, but the language, procedures, and decision theories are virtually the same. Every project has a project manager, superintendent, estimator, and project engineering function or component—whether or not the particular people realize that that is what they are doing. The specific mix of responsibilities will have more to do with the individual's disposition, talent, and experience than with any formal job description.

The manual is to be used by all managers, executives, and professionals in the company. It should be referred to each time any question as to appropriate procedure arises and to confirm one's understanding of the synergy of one activity with all others. Discussions, procedures, letters, and forms have been designed to accommodate the most effective approach to the respective issue, and each has been coordinated with all the related company activities. At any given moment, at least some of this synergy may not be immediately apparent. It is therefore important that the steps described, information generated, and procedure for distribution of result be followed as described. If for any reason your experience begins to suggest an alternative approach, re-read Section 1.7 on quality, and proceed to have the requirements officially changed. Have those changes communicated to all those within the company who will potentially be affected by them.

### 2.2.2 Procedure

The subject of each major section is a distinct discipline within that area of construction management. They're grouped in a way that will facilitate the manner in which the various functions must be performed either because of operating practicalities or by historical arrangement.

As its own discipline, each section can be considered as an individual manual for the specialist. Its relations to the other disciplines are carefully coordinated and interfaced. In this way, for example, the "Purchasing" section of the manual will detail the specific procedures and recommendations for procuring a subcontract for a particular situation, and thereby provide the specific "how-to" for the Purchasing Agent and his or her staff. It will then go on to relate the purchasing efforts to the original estimating function—which was the source of the original purchasing budget in the first place. The Project Engineer and Site Superintendent will be interested not only in their own specialist disciplines but also in how the purchasing effort handled both boilerplate and specific contract requirements throughout the subcontract terms and conditions that they'll have to live with for the project's duration. And so every action that may appear far removed from another person's specific and immediate operating problem can actually be affecting it in dramatic and persistent ways.

Each functional area within the company should have its own copy of procedures. It is not sufficient to have one copy in the company "library," however active that bookcase may be. The procedures must be immediately and simultaneously available to everyone as each person needs them. Each jobsite must have at least one copy, depending upon the actual size of the field staff. Compliance with the stated procedures must be a primary objective of every company member, and the enforcement by every person with any supervisory capacity must be considered to be a primary responsibility.

## 2.3 Correspondence

### 2.3.1 Objective

All correspondence must first be consistent within each project, and then throughout the company. Each communication must be clear, concise, efficient, and made in a way that will make it easy to support and to correlate with the complexities of the project record. Form must be closely coordinated with company routine and special file structures. Information must be factual, opinions must be minimal, and expressions must be controlled.

Correspondence design must begin with the idea that its purpose is to *document*—to identify an issue, create a paper trail, cause action, and prove history. Its success in achieving these targets will directly relate to the success of resolving a given issue. Its efficiency in meeting these goals will directly relate to the time needed to arrive—and therefore the *cost* of arriving—at the end result. The speed with which all information relating to a claim issue can be assembled, for example, will go directly to the time needed for research and preparation (legal fees). The orderliness and comprehensiveness of the information will go directly to effectiveness of any analysis and minimization of possibility of lost information and lost opportunity.

Each time that a correspondence is composed it must be done with the idea that when the item is researched, *all* required pieces of information will be either physically present in the respective file, or that clear references to related information (and their specific locations) are included. There should never be a need to recall circumstances without the benefit of an efficient paper trail.

### 2.3.2 Rules of Effective Project Correspondence

Correspondence is divided between routine, formalized processes (forms, form letters, logs, etc.) and specific communications. The former is treated in great detail throughout the manual. Each such communication has been designed to take advantage of the ideas of this section. The latter are generated daily by nearly everyone in the organization, and must be subjected to the same rules of content, correlation, and efficiency.

The first objective of any correspondence is to communicate information. The final objective is to cause an action. Following the rules below will have a dramatic effect on guiding the outcome:

1. *Keep each project separate.* Never combine discussions of two different projects. Even if it is the same issue ("account status," for example), keep them apart. The reason is to keep future developments focused and efficient. If on a combined letter, for example, the issue is quickly resolved on one project but remains on the other, every discussion will remain unnecessarily complicated. If an owner representative, design professional, or anyone else should be on distribution or should by some other means receive a copy, information regarding other projects is needlessly displayed. Finally, filing becomes cumbersome, and more prone to error.
2. *Confine the subject to a single issue or to a small group of closely or logically related issues.* Keep two or more unrelated items in separate letters. Separate items allow direct focus and quick understanding of the issue. The result will have a better chance of being a faster, more direct response.

   In contrast, a letter with several unrelated items may initially confuse or otherwise distract its reader. Its subjects must be individually researched, and the separate reactions must be piled into an equally confusing response. Both your letter and its response, each containing multiple unrelated items must then be filed anyway in each of the files corresponding to each unrelated issue. So the filing effort must be introduced into the process anyway, but the risk of filing not being properly done (and therefore some opportunity lost later) increases significantly.

   Finally, the worst condition will involve documents containing multiple examples of unrelated information that wind up as evidence in a litigation or arbitration for one of the items. At the very least, expensive time will be spent explaining why all those unrelated items do not apply to the discussion at hand. At worst, you may find yourself having to introduce information that you otherwise would have preferred to leave out of the particular discussion.
3. *Confine letters to a single page if at all possible.* This is actually possible except in the most rare circumstances. Consider the attention span of busy individuals who must understand and respond to your letter. Consider your own reaction when you receive a long, laborious letter, and the work that you've got to go through just to make sure that you've gotten the main point. Make your point(s) in your own letters clear. Get to the point fast. If the issue

is complex, outline the basis in a cover letter that draws the conclusion; attach the support for your contentions in an organized backup package.

4. *Avoid redundancy.* If your letter format includes a clear project reference and subject description, there is no need to waste additional space and attention on repeating the same information in a first paragraph.

5. *Use outline form.* If a long, complicated letter is necessary, do your best to break it down into manageable (understandable) parts. Short letters that contain multiple items should also be treated this way. If the letter is a chronological development, for example, itemize it by date or by event. Clearly identify each individual item. If you're confirming several facts, separate them out of the body copy and number them as separate paragraphs. If you're drawing several conclusions requiring two or more specific actions, reconsider the entire content. Try to break them into individual components, list each item on its own line, and number them. You may discover that it will be more appropriate to construct two or more letters from the information.

   This procedure will have first clarified your own thought process. From there, your representation of the facts will be much more understandable by everyone who will sooner or later become involved. Finally, later research by you or by others will be greatly speeded, by allowing attention to be focused on the points without the need to wade through large amounts of extraneous information.

6. *Keep each discussion simple.* The corollary to the "KISS" formula is that the absolute value of total understanding drops dramatically as the number of people who must understand the issue rises. If ten people must be involved in any given issue, I guarantee that at least three will completely misunderstand it in their first response.

   The second idea is that the higher up any individual (final decision maker) is in his or her organization, the lower your issue in terms of cost and organization impact, the more distracted the decision maker will be. Make it as easy as possible for that individual to see the point, and to see that it is backed up (translation: "it will be a lot of work for me and my organization to even argue with you, much less prevail...").

7. *Use cause-effect style.* Arrange the simplified discussions of your outline presentation in a logical sequence that arrives at its conclusion. Don't jump around. Spend the time to reconsider your draft to be sure that there isn't a more direct route to the conclusion.

8. *Stay factual.* A letter is no place to display emotion or innuendo. In most cases, it is also no place for a stated opinion. Represent names, dates, specific statements, and any other *relevant facts.* Leave out "facts" that really aren't. Don't question the other party's intentions, make accusations, or speculate. All that may have a place in verbal discussions, but not in a bombshell that will be distributed around the planet to people with wide variations in understanding and dispositions—all without you there to explain it. Picture yourself one week from now sitting in a room with several people reviewing your written statements—how comfortable do you feel?

9. *Stay cool.* It's not unusual to receive a letter or otherwise be subjected to something that just sets you off. Anger, insult, disgust, or any other emotion will immediately prompt you to fire off a hot return volley. The unfortunate truth, however, is that such a response will appear short-sighted and unprofessional. At worst, it will blurt out things that you later will regret. Somewhere

in the middle, it will anger at least one of your "opponents," and add time to the resolution of the issue.

In almost all cases, such a response will never resolve the problem. You may get some degree of satisfaction from having "told them," but you will have shoved them a little farther away before you realize that now you've got to start bringing them back.

If you feel a need for such a response, by all means write it. Get it down in all its agonizing detail, but *don't send it.* Let it cool. Put it in your top drawer (get it off your desk in case some well-meaning associate sends it for you). Leave it there for a day. If tomorrow it looks as good as it does at the moment, send it. My bet, though, is that you'll see that it is too long, complicated, redundant, confusing, and maybe insulting. Rewrite the letter focusing on *resolving the problem.* File the first one; it probably does contain a number of useful pieces of information that may become helpful later.

10. *Get to the decision maker.* Most often, the formal route of your correspondence will have long since been determined by your contract working procedure; all letters will be addressed to a specific individual, with copies to certain others. If the highest authority (...the commissioner...) is the addressee, the routine may provide that one of the cc's (the Clerk of the Works) actually will deal with the letter first. If the addressee is the Clerk, the higher authority will be one of the cc's.

    If a shorter time to the decision is needed, or if the overturning of a lower level decision is desired, find a way to lift your letter out of the "routine correspondence" category and get the decision maker's attention directly. The contract itself may turn out to be the greatest help here. The dispute resolution clause may, for example, contain a procedure that provides that "... the commissioner will interpret the plans...." Use the clause to make the commissioner look at your problem if you can't get results at lower levels.

11. *Respect your contract.* Too many times people create or respond to an issue because of assumptions made in the relationships. At the jobsite level, every job "feels" like most others administratively because each will have an architect, owner representative, and construction force. People will accordingly have a great tendency to deal with a problem the way it was done on the last project.

    GC, CM, CM with a GMP, and Design-Build forms of contract will, however, alter the specific relationships. The General Conditions of the Contract will be customized by the Supplementary General Conditions. The entire process will (supposedly) be orchestrated by the Working Procedure section of the specifications. Add to all this state statutes and federal public contract procedures, and it becomes easier to see that you'd better be aware of *your* set of rules in *this* instance. Don't risk having your elegant presentation being dropped in the circular file, or the embarrassment of having to retract a position for being told that "you should have known..." because you *should* have known.

12. *Guide the decision.* Don't simply state your problem. Just stating your problem and asking for an answer isn't any better. Stating your problem and asking for the answer that you prefer may seem better, but it isn't—they already expected that you wanted that answer.

    Your job is to state the problem and to give the technical reasons why the answer must be as you see it. From there, all you should need is a confirmation of the proper interpretation from the proper authority.

Don't, for example, state:

"Both the steel and masonry subcontractors take exception to providing the welded portion of the masonry anchors. Who should do the work?"

Instead, indicate after the first sentence that:

1. Section 04200 3.3a provides that the mason include the fixed portion of all anchors.
2. Detail 6/S3 on the structural drawings shows the welded portion of the anchor on the columns.
3. General Conditions Paragraph 12.3 states that specifications take precedence over the plans.

   Therefore, in accordance with General Conditions Article 3 (the clause that says that *you're* the one who's stuck with the decision) please confirm that the work is included in Section 04200."

13. *Require specific action by a specific date.* After you've guided the action toward the decision you need, state the specific decision as simply and clearly as possible. After that, state the absolute date by which the action is required and pin on some kind of consequence of failing to meet that date.

    Effective remarks may include:

    - "Response after that date will interfere with the masonry work and delay the project."
    - "No action by that date will have allowed the contract work to progress beyond the work in question. In that event, cost and time needed to rework the area will be added to the cost of this additional work."
    - "Response after that date will cause the steel to be delivered to the site without the subject anchors. Even if the work is determined to be contractually theirs, they will be entitled to a change order to cover the extra cost of providing the anchors in the field, instead of fabricating them in their shop."

    Whatever is said, *be accurate.* Don't just use idle remarks or threats that you either cannot or will not be able to follow through on. Don't risk losing your stature as a respected professional over this.

    If the date passes without the required action, react fast. Send a notice, second request, or whatever to prompt the decision. Because your original remarks were accurate, include in your follow-up a notice that the event has occurred; better get moving now before it gets worse. Say something like:

    "As of today, we have no response to my letter of (date). The subject anchors have accordingly not yet been ordered, and the masonry work is now delayed. The total effects on the schedule will be analyzed after your response is finally received. You will be advised at that time of related extra costs."

### 2.3.3 Correspondence Distribution

All correspondence must follow the procedure as outlined in the respective contract Working Procedure or as will be established in the preconstruction or first job meeting. These instructions will incorporate the addressee (architect or owner representative) along with routine copy distribution.

Whether or not instructed to do so officially, other distribution is routinely necessary. This includes:

1. *The architect.* Copies of all correspondence to the owner must be sent to the architect.
2. *The owner representative.* Copies of all correspondence to the architect must be sent to the owner representative. Each must be made aware of every issue on a current basis, and each must be made aware that the other has been so put on notice.
3. *Each person definitely or potentially involved.* If, for example, you are requesting an owner interpretation of a specification potentially involving the steel and masonry subcontractors, those individuals must be copied. If you document an action or statement made by an individual, that person must be put on distribution. This is not only good business but serves as an important notification function. Beyond that, it gives a healthy amount of legitimacy to your statements by showing the world immediately that you're not afraid of holding your remarks up to the light.
4. *File instruction.* The writer of the letter—not the secretary or file clerk—knows all files potentially affected by the issue and its document. At the time the letter is written, decide all the files that must receive a copy, and indicate such in a final "cc: File: x, y, z," where x, y, and z are specific file folders. In this way, anyone can be given the filing task without the risk of misfiling.

### 2.3.4 Correspondence Checklist and Desk Display *(page 2.9)*

The checklist below summarizes the objectives of Section 2.3.2 as a reminder to apply the principles in *every* situation. Photocopy and cut out the Checklist. Either post it near your work area, or fold on the perforation and stand it on your desk. Keep it as a constant reminder, refer to it and keep your correspondence focused on achieving timely objectives—not on keeping fires burning.

## 2.4 Files and File Management

### 2.4.1 Overall File Structure

The focus of this manual is on project management. The central accounting files of contracts, subcontracts, purchase orders, accounts payable, accounts receivable, job cost, general ledger, and tax applications are beyond the scope addressed here. Discussion of accounting will be detailed as necessary for comprehensive treatment by the project management function.

Accordingly, the project management files for each project will consist of:

The Contract Documents

The General Project File

Duplicate "Correspondence File" Books

Clarifications/Changes Log and Books

Subcontractor Summary and Phone Log

Jobsite Subcontractor Performance Summary and Phone Log

Submittal Log

2.3.4

**Correspondence Checklist & Desk Display**

(Stand Backing)

(Fold)

**Correspondence**

- Keep each project separate.
- Use single subject, or small group.
- Use single page, if at all possible.
- Avoid redundancy.
- Use outline form.
- Keep each discussion simple.
- Use cause-effect style.
- Stay factual.
- Get to the decision maker.
- Require specific action by a specific date.
- Guide the decision.

### 2.4.2 The Contract Documents *(page 2.10)*

The complete set of Contract Documents includes the list on page 2.10.

The Bid Document and Agreement components of the Contract will be filed and managed with the General Project File. Here, we'll begin with the setup and maintenance of the Plans, Specifications, and Changes.

### 2.4.3 Setup and Maintenance of the Plans, Specifications, and Changes

1. Initial Setup

   *a. Consolidation of Drawings*

   (1) Arrange a single set of plans marked "Office" at the central office on a plan stick and set in a rack. Separate major plan divisions of large sets on individual sticks.

**The Contract Documents**

## Bid Documents

- Instructions to Bidders
- Prime Contract Bid
- Subcontract Bids

## The Agreement

- Owner / Prime Contractor
- Prime Contractor / Sub-Vendors
- Changes executed after the original agreement

## Plans

- Architectural
- Site
- Structural
- Fire Protection
- HVAC
- Electrical
- Specialty- (ies)
- Shop Drawings
- Clarifications
- Changes

## Specifications

- General Conditions
- Supplementary General Conditions
- Working Procedure
- Technical Specifications
- Referenced Technical Standards
    - Building & other codes
    - State & federal regulations
    - Labor standards
    - Environmental standards

(2) Arrange *two* duplicate sets for the jobsite; one clearly marked "Jobsite," and the other marked "As-Builts."

(3) Place three complete copies of specifications in hard-cover three ring binders, each labeled and distributed with the respective matching set of plans. Select a binder color that will later be matched with the file tabs and all other binders.

(4) The "Office" and "Jobsite" copies will be managed from here in a way that will help coordinate all relevant information properly during construction. The "As-Built" set will develop over the life of the project as a complete record of all actual work to be turned over to the owner at the end of the project.

*b. Post Each Addendum*

(1) On both the "Office" and "Jobsite" sets of plans and specifications, physically cut out each addendum note and glue or tape at the area of plan or

specification altered by it. Do not cover or otherwise hide the original requirement, but add a clear reference to the addendum change. Next to each such posting, note which addendum number it came from. Short notes or minor changes can be made directly in red pen, with the appropriate addendum reference. It will be sufficient on the "As-Built" set to simply attach a copy of all addenda for the record.

(2) The posting of each individual addendum remark in each's correct location *is absolutely essential to the avoidance of serious and costly problems* resulting from simple oversight of officially changed conditions. The method will eliminate the need throughout the project's life for constant hunting through an entire set of addenda *every* time *any* specification or plan note is referenced to be sure that it is the correct one.

2. Maintenance

*a. Post Changes*

(1) As the job progresses, there will be numerous sources of modifications to the physical construction. Change orders, clarifications, job meeting discussions, and even telephone conversations will officially alter the requirements.

Each of these changes will have its own development history and paper trail to varying degrees of complexity, and its proper documentation is covered in detail in the respective sections of this manual.

(2) The important idea here is simply to be certain that every modification—however simple or complex—is clearly marked in color on both the "Jobsite" and "Office" sets, with appropriate references to the correct source of the official change. Forms of these alterations may include:

- SK-Drawings and other clarification sketches
- Job meeting discussions
- Telephone and other conversations

(3) *"SK" drawings or other clarification sketches* will be added. Clarification sketches of construction details not involving changes in cost or time are a matter of course. Highlight all areas affected on the plans, referring to the new sketch. Tape a photocopy of the change on the plan adjacent to the area if there is room. If there is no room, tape the sketch on the opposite (left) sheet of the plans (back of the previous plan page). At the highlighted area of the original plan, include the note to "Refer to SK-7 taped opposite."

Many times, changes and clarifications that develop during discussions tend to become more complicated than may have originally been expected. In these cases, many designers resist taking the time to make sketches that properly accommodate the changes. If anyone involved with proper coordination (starting with yourself) is at all confused with change descriptions, or if you feel that there is any potential at all for confusion now or later, *insist* that the designer properly issue such a complete *written* clarification, and treat as described above.

Follow the instructions in Section 2.4.5.

*Clarification / Change Log* to keep a chronological record of all changes as they may develop.

*b. Job Meeting Discussions*

Job meetings and their minutes are important coordination and documentation vehicles, not only of project effects, but of their chronologies and responsibilities. If conducted and documented as described in that section of the manual, the details will be clear and properly correlated.

The logistical difficulty in the timely incorporation of the individual items lies in the sheer number of them. As the number of items inevitably increases, the risk of losing one or a few before it's too late rises dramatically.

Most job meetings are held at the jobsite. If not, get the location changed to the jobsite. Accordingly, have the "Jobsite" set of plans and specifications available at every job meeting. Immediately upon the resolution of any item that results in a change to the documents, mark the set right at the meeting, noting the job meeting number and date as the source of the change. It will be an easy reference that will be supported by the respective meeting minutes if it is later questioned, and you will have completely eliminated all possibility that the change would be forgotten.

As meeting attendees continually observe this procedure as a matter of course, they will, in turn become more comfortable in the legitimacy of the notations on the official project documents because they will for the most part have witnessed how the notes were made.

c. *Telephone and Other Conversations*

Apparently minor changes and clarifications are a daily routine ("...there's no dimension..."). Simple conversation with an owner representative or design professional gets a fast answer that is acted on immediately:

*Question:* ... I'm laying out the wall now; what's the dimension?

*Answer:* The owner's furniture is 3′ wide; make it 3′3″.

Note the new information directly on the plans, indicate the source of the information *by name,* and date it. Ideally, have that individual sign or initial the remark right on the drawings. At the very least, immediately write a confirming memo, and send a copy to be so noted on the "Office" set of plans.

If the instructions get complex or you feel may otherwise result in confusion now or later, ins0ist on a sketch and treat it as described in (1) above.

## 2.4.4 General Project File

Immediately upon notice of award of a contract, the project manager will direct the assembly of the General Project File. There will be *two nearly identical files maintained*—one at the jobsite and one at the central office. The central office file will be considered to be the primary file and the jobsite file considered to be the copy. As such, the integrity of the central office file must never be compromised. The jobsite file may have sensitive documents omitted.

Coordinate the label color with the specification binder of 2.4.2 that has distinguished the new file sets from all other jobs. All files as developed, file books, and log books will be similarly color coded for ease of reference.

The General Project File will be divided between Project Administration and Technical Construction. The individual folders will be titled and arranged as follows:

**Project administration**

- General Contract
- Job Cost Summary and Estimate Details

- Cost Progress Reports
- Daily Field Reports
- General Payment and Performance Bonds
- General Insurance Policies and Certificates
- Subcontractor Insurance Certificates
- General and Subcontractor Schedule of Values
- General Payment Applications
- Purchase/Award Schedule
- Building and Special Permits
- Certificate of Occupancy/Substantial Completion
- Punchlist/Final Completion
- Subcontractor Approvals (if required)
- Progress Photographs
- Monthly Narratives (if used)
- Jobsite Utilization Program
- Guarantees/Warranties
- As-Built Documents
- Maintenance and Operating Manuals
- General Notifications (to all subs & suppliers)
- Baseline Schedule
- Periodic Schedule Revisions
- Job Meeting Minutes
- Special Meeting Minutes
- Change Order Summary
- Individual Change Order Files
- Backcharge Summary
- Potential Owner Claim Summary
- Potential Subcontractor Claim Summary
- All others as appropriate; refer to the General Conditions and Supplemental General Conditions as sources of special requirements.

#### Technical construction divisions

File folders for Divisions 2 through 16 of the technical specifications will be keyed to directly correlate with specific specification division and section numbers. Within each technical division, divide the files to correlate with the specific arrangements of the respective bid packages. Never combine subcontracts or bid packages in the same folder "for convenience." It will lead to a lot more work and confusion later.

**Example:** "Division 2: Sitework" will likely contain a major site bid package that may include 80 percent of the work described in the division, thereby incorporating several site specification sections. The landscaper, however, will be a separate subcontractor, and other items, such as playground equipment may also be purchased from separate vendors. Each separate vendor within Division 2 must accordingly have its own file appropriately headed.

#### Bid package file organization

Each of the separate bid package files as described above will be physically divided between *Contract and Correspondence* and *Submittals.* Submittals and shop drawings tend to become bulky and become a physical inconvenience when handling nonsubmittal material. A separate manila folder (or set of folders as may become necessary) placed within hanging folders identified as the same bid package and placed adjacent to the Contract and Correspondence folder will usually be sufficient.

#### Subfiles

Within the Contract and Correspondence file of a bid package, any issue generating at least two pieces of correspondence with the potential of generating more will justify the preparation of a Subfile. This is simply an additional manila folder appropriately headed, and placed in the bid package hanging folder adjacent to the Contract and Correspondence file folder.

**Example Subfile Folder Heading**

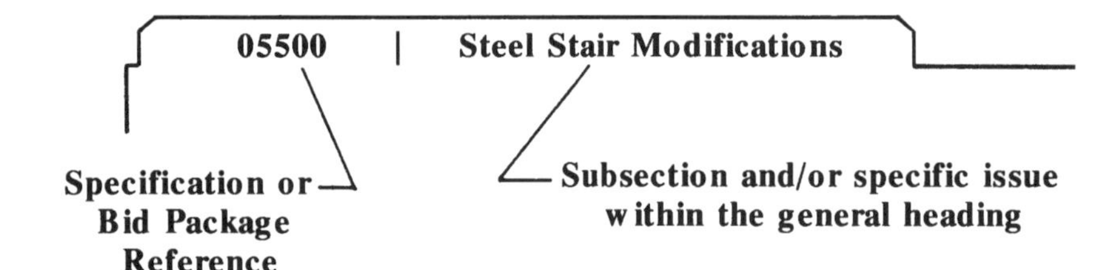


#### Duplicate "Correspondence File" books

At the same time the General Project File is set up, prepare Book #1 of the *Correspondence File.* This is a 2″ or 3″ hardcover, three-ring binder in a color that will match that selected for the file folders and specification binder, identified as follows:

- Project Name
- Owner Project Number
- Company Project Number
- Book Number
- Beginning Date
- Reserved area to insert the end date, upon preparation of the next book in chronological order.

Items are to be filed in the Correspondence Book in reverse chronological order (most recent up front) per the date of the correspondence (not the date of receipt), and include:

*a.* Duplicate copies of all letters to and from the Company and to any other party regarding the project in any way (copies of which are received by the Company).

- Stamp all letters with the date of receipt, or mark the date of receipt in the upper right corner and initial it.

*b*. Copies of all transmittals (in and out) regarding everything but shop drawings or other approval submittals. (Refer to *Transmittal Procedure and Use*). Copies of transmittals regarding approval submittals will be filed in that binder as described below.

Attach the architect/engineer's transmittal to the back of the Company transmittal. (Refer to *Shop Drawing Summary Record Procedure*).

Be sure that each item is marked with the proper *file instruction*. For Transmittals, refer to *Transmittal Procedure and Use.* All Company correspondence should already contain such a file instruction. If not, put it on now. Add such instruction as described in Section 2.3.3 on all correspondences—usually in the upper or lower right sides of the correspondence.

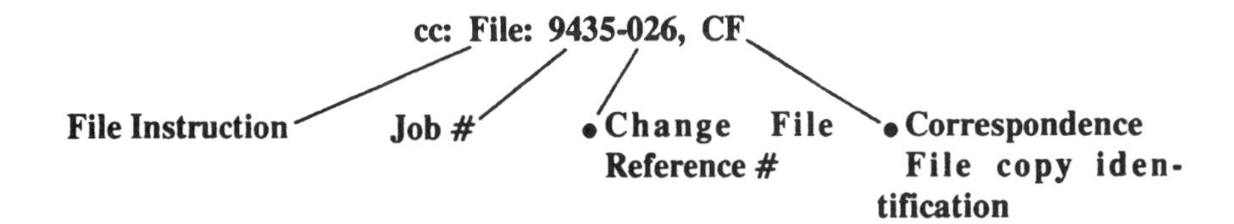


## 2.4.5 Clarification/Change Log

Prepare a 2" or 3" three ring binder (color to match the project set) labeled as *Clarifications / Changes.* Use tab inserts to separate the log book into three major divisions, labeled as:

1. Clarifications/RFI's (Requests for Information)
2. Change Order Summary
3. Clarification Sketches

Insert the RFI Summary Log form provided in Section 4.12 of the manual in the first section, and the Change Order Summary Log form provided in Section 4.9 in the second section. The third section is reserved to file copies of the actual transmittals, clarification sketches, requests for information (RFI's) and their actual responses—all in strict chronological order. Add additional binders as needed for this section three. The two log sections will always remain in the front of Book #1.

## 2.4.6 Subcontractor Summary and Phone Log

For use in the central office (or where the central project engineering function will consistently be performed), prepare a 1" three ring binder (color to match the project set) labeled as *Subcontractor Summary.* Insert a set of alphabetically tabbed dividers that will later receive the *Subcontractor / Supplier Summary* form described in Section 4.4, followed by the *Subcontractor/Supplier Reference Form* (Section 4.5) and the *Phone Log* form (Section 4.4) for each respective vendor as they are signed on to the project.

### 2.4.7 Jobsite Subcontractor Performance Summary and Phone Log

For use at the jobsite, prepare a 1" three ring binder (color to match the project set) labeled as *Jobsite Subcontractor Log.* Insert a set of alphabetically tabbed dividers that will later receive the *Subcontractor/Supplier Jobsite Performance Form* described in Section 4.19, followed by copies of the same *Subcontractor/Supplier Reference Form* (Section 4.5), and the *Phone Log* form (Section 4.4) for each respective vendor as they are signed on to the project.

Insert starting sets of (30) sheets of the Performance Form and (600) sheets of the *Phone Log* form to be used as the project develops. Copies of the *Reference Form* will be forwarded to the jobsite as they are received completed by the respective vendors.

### 2.4.8 Submittal Log

For use at the central project engineering location, prepare a 1" three ring binder (color to match the project set) labeled as the *Submittal Log.* Insert a starting set of (30) sheets of the *Submittal Log Form* as described in Section 4.9. Each log sheet will be prepared and maintained as each vendor is signed on to the project in accordance with the instructions of Section 4.9 of the manual.

## 2.5 How to Recover a Mailed Letter

If you ever need to, it's useful to know that you *can* recover a letter after it's been mailed. The correct form to be used is available at any time from any post office.

Ask for PS Form 1509.

The success of its use, of course, depends upon how soon the recovery is implemented after the document to be retrieved had originally been mailed.

## 2.6 Field Labor Time Reporting

### 2.6.1 General

Accurately reporting field labor is of course necessary to process the respective payroll correctly. The other objective is to substantiate actual costs as they are applied to the individual project components. It is important that the forms be completed throughout each day, as the information is being generated. Anything less than this type of attention may compromise the accuracy of the information.

### 2.6.2 Field Payroll Report Form

The Field Payroll Report Form is broken into the specific activities that are performed on a given day by a particular employee. A numbering system will be assigned to the possible labor categories as determined by the project work estimates. If there is no current activity number for a new item of work being done, contact the project engineer to secure a new added number assignment.

The labor hours are reported against those activity numbers. The result will be a job cost report that will identify the exact labor cost actually applied to the individual job components.

This kind of detail will provide valuable comparisons to the cost estimates for the respective activities. Actual costs will be tracked against the estimates as they occur, in order to allow identification of potential problems before opportunities for correction have passed.

Finally, the accurate detail of the records will provide an indisputable account of the actual cost of time-and-material, change order, or any matter in dispute. In each of these cases, absolute attention must be devoted to timeliness and accuracy of the information.

## 2.6.3 Procedure

1. Each hourly wage individual on the project staff is required to prepare and maintain the forms on a daily basis. *There are no exceptions. All* salaried field personnel are required to use the *Weekly Administrative Time Sheet* as described in Section 2.9 of the Manual.
2. Most field personnel are assigned to a single project. For those with any multi-project responsibility, a single form (or sets of forms) must be completed for *each individual project.* Never combine projects on the same form for any reason.
3. Treat change orders, potential change orders, work done under protest, and work that is or may be the subject of a claim as a separate project. The general project number with the change suffix will serve as the complete "Job Number."

**2.6.3**

**Example Change File Identification**

4. The forms should be filled out throughout the day as the activities are changing, but in any case they *must* be filled out *every day* at an absolute minimum.
5. Use the *Project Administrative Activity List* included in Section 2.8 of the Manual as the source for the activity numbers to be used to identify the particular administrative activities.
6. To report activity information for specific field work (concrete formwork, rough carpentry, etc.) secure the general activity list from the central office as determined for the specific project.

   This is a listing of each trade activity to be performed by Company personnel that has been prepared from the original bid package estimate for the work category. Refer to Sections 4 and 8 of the Manual for details on its preparation.

7. All completed forms are to be returned to the central office each week on the designated day. Check with your Accounting Department for the requirement.

### 2.6.4 Sample Field Payroll Report Form—Completed Example *(page 2.19)*

### 2.6.5 Sample Field Payroll Report Form—Blank Form *(page 2.20)*

## 2.7 Administrative Time Reporting

### 2.7.1 General

Because Project Managers, Project Engineers, Project Executives, Estimators, and Administrative Assistants are salaried personnel, it is not necessary to report their labor hours for payroll. The purpose of properly reporting administrative time is to provide management with accurate information regarding:

- Where administrative overhead is spent
- Where the largest percentage of overhead is spent in terms of the project's life cycle
- What percentage of administrative overhead is spent in the field and in the central office
- What charges to apply to changes, delays, interruptions, accelerations, and reschedules
- What are the actual costs of the project's general conditions components

What is necessary is for the Company to be able to determine on a current basis the divisions of time spent by salaried personnel on the various planned and other activities. Reasons for this include:

1. For planning reasons, a knowledge of where time is really being spent on routine operations of the individuals is helpful for determining future time/cost requirements for particular assignments.
2. Of particular interest from that point will be the time soaked up by changes, problems, or on other activities that are otherwise not the direct responsibility of company individuals to perform directly (such as managing a subcontractor's subcontractor).

   In all these cases, it is crucial to maintain accurate record of *all* costs and efforts related to these changed conditions if the company is to retain any ability to recoup these extra costs and eventually remain whole.
3. Even in the cases of central office personnel such as project executives, estimators, accountants, and so on who divide their time simply by project and not by specific activity within each project, the information will be extremely useful if it becomes necessary to display the disproportionate amounts of time being spent by these individuals in problem situations.

**2.6.4**
**Sample Field Payroll Report Form**
**(Completed Example)**

## FIELD PAYROLL REPORT

PROJECT: Plainville Library NO: 9415 Week Ending: 1 / 22 / 94

| EMPLOYEE | SUN Act No | SUN Hrs | MON Act No | MON Hrs | TUE Act No | TUE Hrs | WED Act No | WED Hrs | THUR Act No | THUR Hrs | FRI Act No | FRI Hrs | SAT Act No | SAT Hrs | TOTAL HOURS |
|---|---|---|---|---|---|---|---|---|---|---|---|---|---|---|---|
| No. 5467 | | | 0151 | 4 | 0151 | 4 | 0151 | 3 | 0151 | 3 | | | | | 14 |
| | | | 0155 | 3 | 0155 | 4 | 0155 | 3 | | | 0155 | 4 | | | 14 |
| Name: | | | 0190 | 1 | | | 0190 | 2 | 0190 | 3 | 0190 | 4 | | | 10 |
| | | | | | | | | | 0210 | 2 | | | | | 2 |
| ED FREDRICK | | | | | | | | | | | | | | | |
| | Total | | Total | 8 | Total | 8 | Total | 8 | Total | 8 | Total | 8 | Total | | 40 |
| No. 3320 | | | 0180 | 2 | 0180 | 4 | 0180 | 4 | | | 0180 | 2 | 0810 | 2 | 14 |
| | | | 0110 | 2 | | | | | 0110 | 4 | 0110 | 2 | | | 8 |
| Name: | | | 1900 | 4 | 1900 | 4 | 1900 | 2 | 1900 | 2 | | | | | 12 |
| | | | | | | | 1400 | 2 | | | | | | | 2 |
| M. LYDEN | | | | | | | | | 1550 | 2 | 1550 | 4 | | | 6 |
| | Total | | Total | 8 | Total | 8 | Total | 8 | Total | 8 | Total | 8 | Total | 2 | 42 |
| No. 4141 | | | 2200 | 4 | 2200 | 4 | 2200 | 4 | 2200 | 2 | 2200 | 3 | | | 17 |
| | | | 2210 | 4 | | | | | | | 2210 | 3 | | | 7 |
| Name: | | | 2100 | 2 | 2100 | 4 | | | 2100 | 2 | 2100 | 2 | | | 10 |
| | | | | | 3550 | 2 | 3550 | 2 | | | | | | | 4 |
| L. TEE | | | | | | | 3500 | 2 | 3500 | 4 | | | | | 6 |
| | Total | | Total | 10 | Total | 10 | Total | 8 | Total | 8 | Total | 8 | Total | | 44 |
| No. | | | | | | | | | | | | | | | |
| Name: | | | | | | | | | | | | | | | |
| | Total | | Total | | Total | | Total | | Total | | Total | | Total | | |
| No. | | | | | | | | | | | | | | | |
| Name: | | | | | | | | | | | | | | | |
| | Total | | Total | | Total | | Total | | Total | | Total | | Total | | |

## 2.6.5
## Sample Field Payroll Report Form

### FIELD PAYROLL REPORT

PROJECT: ______________________ NO: ____________ Week Ending: ____/____/____

| EMPLOYEE | SUN | | MON | | TUE | | WED | | THUR | | FRI | | SAT | | TOTAL HOURS |
|---|---|---|---|---|---|---|---|---|---|---|---|---|---|---|---|
| | Act No | Hrs | Act No | Hrs | Act No | Hrs | Act No | Hrs | Act No | Hrs | Act No | Hrs | Act No | Hrs | |
| No. ____ Name: ____ | Total | | Total | | Total | | Total | | Total | | Total | | Total | | |
| No. ____ Name: ____ | Total | | Total | | Total | | Total | | Total | | Total | | Total | | |
| No. ____ Name: ____ | Total | | Total | | Total | | Total | | Total | | Total | | Total | | |
| No. ____ Name: ____ | Total | | Total | | Total | | Total | | Total | | Total | | Total | | |
| No. ____ Name: ____ | Total | | Total | | Total | | Total | | Total | | Total | | Total | | |

To treat these individuals' time reporting in a more simplified (but more generalized) manner, their records can be maintained on a simplified project-by-project basis and consolidated monthly.

### 2.7.2 Procedure

1. The *Weekly Administrative Time Sheet* is to be kept by *all* salaried jobsite personnel, and by any other individuals so designated by Company procedure to report their project time split into the individual activities.

   Site superintendents will generally have single-project responsibility. All off-site and central office positions will generally have multi-project responsibilities to varying degrees.
2. Throughout and at the end of each day, general supervisory people with multi-project responsibility, all central office staff (estimators, purchasing, etc.) list each project worked on at the left side of the form. An estimate of the amount of time spent on each activity within each project is included in the appropriate day column.

   Use the *Project Administrative Activity List* of Section 2.8 of the Manual as the starting listing of activity number to be assigned on the form. Any additional activities and number listing as may become necessary will be determined by project engineering as coordinated with project accounting per the requirements of Sections 4 and 8 of the Manual.
3. Treat changes orders, potential change orders, work done under protest, and work that is or may be the subject of a claim as a *separate project.* The general project number with the change suffix will serve as the complete "Job Number."

**Example:** 9325-042

General Project Number | Change File Number

4. For those individuals who are authorized by general management to report their time simply divided by project, use the *Monthly Administrative Time Sheet.* Understand that, although completing this form requires less time and effort on the part of the individual, the respective information has accordingly been greatly generalized. Consider this carefully before the procedure is officially adopted.
5. The form must be turned in to central accounting at the designated day each week (or at the end of each month for the Monthly Form) as a condition for release of the individual payroll check.

### 2.7.3 Sample Weekly Administrative Time Sheet—Completed Example *(page 2.22)*

### 2.7.4 Sample Weekly Administrative Time Sheet—Blank Form *(page 2.23)*

## 2.7.3
## Sample Weekly Administrative Time Sheet
## (Completed Example)

### WEEKLY ADMINISTRATIVE TIME SHEET

Form No. 1

NAME L. LEONARDO MONTH JAN 19 94 WEEK NOS. 1 2 3 4
POSITION ADMIN. ASSISTANT REVIEWED BY SC DATE 1.31.94

(2) WEEK PERIOD (Sunday to Saturday)

| PROJECT NAME | PROJ. NO. | S | M | T | W | T | F | S | S | M | T | W | T | F | S | TOTAL |
|---|---|---|---|---|---|---|---|---|---|---|---|---|---|---|---|---|
| FIRE HOUSE | 9205 | | 2 | 4 | 3 | 2 | 3 | | | 3 | 2 | 4 | 4 | 5 | | 32 |
| AJH CNTR. | 9320 | | 1 | 4 | 3 | 3 | 2 | | | 3 | 3 | 2 | 4 | 3 | | 28 |
| JEFFERSON SCH. | 9330 | | 3 | 0 | 1 | 2 | 1 | | | 1 | 2 | 1 | 0 | 0 | | 11 |
| COMM. SQUARE | 8925 | | 2 | 0 | 1 | 1 | 2 | | | 1 | 1 | 1 | 0 | 0 | | 9 |
| | | | | | | | | | | | | | | | | |
| | | | | | | | | | | | | | | | | |

(2) WEEK PERIOD (Sunday to Saturday)

| PROJECT NAME | PROJ. NO. | S | M | T | W | T | F | S | S | M | T | W | T | F | S | TOTAL |
|---|---|---|---|---|---|---|---|---|---|---|---|---|---|---|---|---|
| FIRE HOUSE | 9205 | | 3 | 2 | 2 | 4 | 4 | | | 2 | 3 | 4 | 3 | 3 | | 30 |
| AJH CNTR. | 9320 | | 3 | 2 | 2 | 1 | 4 | | | 2 | 4 | 4 | 3 | 3 | | 30 |
| JEFFERSON SCH. | 9330 | | 1 | 2 | 3 | 3 | 0 | | | 3 | 1 | 0 | 1 | 1 | | 13 |
| COMM. SQUARE | 8925 | | 1 | 2 | 1 | 0 | 0 | | | 1 | 0 | 0 | 1 | 1 | | 7 |
| | | | | | | | | | | | | | | | | |
| | | | | | | | | | | | | | | | | |

SIGNATURE L. Leonardo

## 2.7.4
## Sample Weekly Administrative Time Sheet

### WEEKLY ADMINISTRATIVE TIME SHEET

Form No. ______

NAME ____________________ MONTH ____________ 19 _____ WEEK NOS. ___ ___ ___ ___
POSITION ____________________ REVIEWED BY ____________ DATE ____________

(2) WEEK PERIOD (Sunday to Saturday)

| PROJECT NAME | PROJ. NO. | S | M | T | W | T | F | S | S | M | T | W | T | F | S | TOTAL |
|---|---|---|---|---|---|---|---|---|---|---|---|---|---|---|---|---|
| | | | | | | | | | | | | | | | | |
| | | | | | | | | | | | | | | | | |
| | | | | | | | | | | | | | | | | |
| | | | | | | | | | | | | | | | | |
| | | | | | | | | | | | | | | | | |
| | | | | | | | | | | | | | | | | |

(2) WEEK PERIOD (Sunday to Saturday)

| PROJECT NAME | PROJ. NO. | S | M | T | W | T | F | S | S | M | T | W | T | F | S | TOTAL |
|---|---|---|---|---|---|---|---|---|---|---|---|---|---|---|---|---|
| | | | | | | | | | | | | | | | | |
| | | | | | | | | | | | | | | | | |
| | | | | | | | | | | | | | | | | |
| | | | | | | | | | | | | | | | | |
| | | | | | | | | | | | | | | | | |
| | | | | | | | | | | | | | | | | |

SIGNATURE ____________________

## 2.8 Project Administrative Activity List

### 2.8.1 General

The Project Administrative Activity List is to be used by all hourly and salaried personnel for their time reporting as described in Sections 2.6 and 2.7 of the Manual. If new categories become necessary to accurately describe all activities, secure new numbers and activity descriptions from the project engineer who will secure the information in accordance with the requirements of Section 4 of the Manual.

Because actual and potential changes are *treated as separate projects,* there is no need to make the distinction within the individual activity descriptions. The activities themselves will apply to the respective contract or changed work as so designated on each form.

### 2.8.2 Project Administrative Activity List

| *Activity Number* | *Activity Description* |
|---|---|
| ______________ | Set up/maintain field office |
| ______________ | As-Built Documents |
| ______________ | Field Studies/Reporting |
| ______________ | Field Engineering & Layout |
| ______________ | Field coordination |
| ______________ | Off-site travel time |
| ______________ | Design coordination/clarification |
| ______________ | Finalize change designs |
| ______________ | Negotiate/finalize change order |
| ______________ | Winter conditions preparation |
| ______________ | Progress schedule generation |
| ______________ | Progress schedule updating and reporting |
| ______________ | Quantity survey estimating |
| ______________ | Cost Estimating |
| ______________ | Secure sub-bid proposals |
| ______________ | Estimate/proposal preparation |
| ______________ | Job Meeting attendance & documentation |
| ______________ | Special Meeting attendance & documentation |
| ______________ | Subcontractor submittal review/processing |
| ______________ | Company submittal review/processing |
| ______________ | Material expediting |
| ______________ | Organize/monitor/document backcharges |
| ______________ | Subcontractor payment review/approval |
| ______________ | Prepare/process general payment application |
| ______________ | Cost progress reviews/documentation |
| ______________ | Staff hiring/interviews |
| ______________ | Exit interviews |
| ______________ | Staff performance evaluations |
| ______________ | Proposal preparation |

| *Activity Number* | *Activity Description* |
|---|---|
| __________ | Proposal resubmission/negotiation |
| __________ | Progress photographs |
| __________ | Field Reporting |
| __________ | Prebid conferences |
| __________ | Scope review conferences |
| __________ | Value engineering |
| __________ | Purchasing |
| __________ | Safety reviews |
| __________ | Accident investigation/reporting |

## 2.9 Expenses and Reimbursements

### 2.9.1 General

It is the policy of the Company to reimburse employees for all normal and necessary expenses incurred in connection with company business. Good expense reporting and tight budget control are as important as efficient production.

### 2.9.2 Allowable Items

1. Fuel, oil, tolls, parking expenses directly related to company business. Use of your personal vehicle must be approved in advance by the Project Manager.
2. Lunch, or other entertainment for customers, design professionals, and owner representatives only if generally approved in advance by the Project Manager.
3. Small procurement for the project in amounts not exceeding $50.00. All small tools, supplies, etc., that have not been completely expended must be immediately turned over to the Company after their use is complete.
4. Material, tool, equipment, or supply procurement for the project in amounts exceeding $50.00 only if approved by the Project Manager in advance.
5. Maintenance items for company issued equipment (batteries and film for cameras, etc.).
6. Overnight travel expenses only with the approval of the Project Executive.

### 2.9.3 Guidelines for Expense Reporting

1. Expense reports are to be submitted in neat form, complete, in ink.
2. Tape appropriate receipts in order to 8 ½″ × 11″ paper, and attach to the respective expense report.
3. Completed Expense Reports are to accompany the Weekly Time Sheet or Administrative Time Sheet (Sections 2.6 and 2.7 of the Manual).

### 2.9.4 Sample Expense Report—Blank Form *(page 2.26)*

## 2.9.4
## Sample Expense Report

TO ____________________ FROM ____________________

| FOR PERIOD ENDING | SUN. CITY | MON. CITY | TUES. CITY | WED. CITY | THUR. CITY | FRI. CITY | SAT. | TOTALS | |
|---|---|---|---|---|---|---|---|---|---|
| 1 HOTEL MOTEL | | | | | | | | | 1 |
| 2 BREAKFAST | | | | | | | | | 2 |
| 3 LUNCH | | | | | | | | | 3 |
| 4 DINNER | | | | | | | | | 4 |
| 5 PLANE-RAIL BUS FARE | | | | | | | | | 5 |
| 6 LOCAL TAXIS BUS FARE | | | | | | | | | 6 |
| 7 AUTO EXPENSE REPAIR-TIRES SUPPLIES | | | | | | | | | 7 |
| 8 GAS AND OIL | | | | | | | | | 8 |
| 9 LUBRICATION AND WASH | | | | | | | | | 9 |
| 10 GARAGE PARKING | | | | | | | | | 10 |
| 11 PHONE TELEGRAMS | | | | | | | | | 11 |
| 12 TIPS | | | | | | | | | 12 |
| 13 TOLLS | | | | | | | | | 13 |
| 14 ENTERTAINMENT | | | | | | | | | 14 |
| 15 | | | | | | | | | 15 |
| 16 | | | | | | | | | 16 |
| TOTALS | | | | | | | | | |

| MILEAGE RECORD | | |
|---|---|---|
| END OF TRIP | | REMARKS: |
| LESS – START | | |
| MILES PER TRIP | | |

I HEREBY CERTIFY THAT THE ABOVE EXPENDITURES REPRESENT CASH SPENT FOR LEGITIMATE COMPANY BUSINESS ONLY AND INCLUDES NO ITEMS OF A PERSONAL NATURE.

SIGNED ____________________

| DATE | REPAYMENT RECAP | AMOUNT | APPROVAL | CASHIERS MEMO |
|---|---|---|---|---|
| | ADVANCE RECEIVED | | | |
| | REIMBURSED | | | CHECK NO. |
| | TOTAL | | | |
| | EXPENSE FOR WEEK | | | |
| | OVER OR SHORT | | | DATE AMOUNT |

Section

# 3

# General Contracts

## 3.1 Section Description

### 3.1.1 General

This section is for use by all company personnel having any project management, project engineering, and site superintendence responsibilities. It should be read through in sequence and then referred to in any time of question.

The section's purpose is to provide an overall perspective on the general nature of (1) contract law and (2) construction contract law. It is not intended to be exhaustive in this regard, so all company members with any project responsibility should make it their business to become completely familiar with all nuances of at least the principles discussed here by using the many good books available on the subjects. Books on construction claims, for example, are very useful not so much for the idea that we intend to take aggressive claims postures at every turn, but in their structural approach to analyzing and strictly applying general contract language in a technically legal sense. From that understanding, we all might better apply our contracts in *practical* situations while *avoiding* lawyers.

It is important to be able to quickly assess the total picture of each situation almost as it is occurring. From that position of complete information, we will all be in a better position to assess all options, and *to visualize each one through to its final conclusion*—even if that conclusion must be arbitration. The better our ability to do this, the better will be our ability to quickly determine appropriate responses for the moment and those to follow. The details will immediately become clear, right down to the correspondence distribution and file procedure.

### 3.1.2 The G.C. as a "Conduit"

As a general or prime contractor, or as a construction manager, it is your responsibility to *coordinate* the work. It is *not* your responsibility to coordinate the design (see Section 3.8), interpret the contract documents (see Section 3.5), to "approve" submittals (see Section 3.6), or through any other mechanism unnecessarily assume any liability that correctly belongs to the architect/engineer or the owner.

General, design-build, and the various forms of construction management contracts all have significant and subtle differences in both their express relationships and the manner in which those relationships are treated. However well meaning, we as builders get into many problems because most projects "feel" like most others in terms of the relationships; there's always an owner, design force, and construction force each doing what pretty much looks like what was done on every other project.

Our problems as contractors distill to two levels:

1. Each contract *is* different. It is *crucial* that we must always be aware of the *specific* language for the current situation on *this* project. We must constantly remind ourselves to check, verify, then check again before proceeding on an item in the way we've proceeded before. You may actually be digging the hole deeper before you start climbing out.
2. We are too often too eager to interpret the documents and to direct subcontractors and suppliers directly as to their required performance under the contract. In the interest of moving the project along, ending what on the surface may seem to be a frivolous action on the part of a sub, or just trying to do what

seems to be the right thing, we unintentionally assume the *liability* for those actions that belongs with the designers and/or the owner.

Read and reread Section 3.3.8, the *Pass-Through Clause.* Understand that each subtrade is bound to the owner for its work to the same extent that you are, and that the owner is bound back to them in the same way. Understand not only that you should not interpret the documents (and thereby assume the responsibility for the interpretation) but that it is not your right to; that's the *owner's* job. Learn to hold the owner's feet to the fire (figuratively) to get the decisions that you're entitled to in order to keep the contractual relationships intact.

Develop this perspective, and hold onto it dearly as you consider the discussions of this and all other sections of the manual.

## 3.2 General Principles of Contracts as Applied to Construction

### 3.2.1 General

There are several basic rules of contract interpretation that generally apply to all contracts, regardless of the industry. These principles form the basis for the initial application of the law in deciding how conflicts will be resolved. Even if these ideas were absent in the form of strict rules, their principles are useful to apply to determine the practical structures of various situations and options. They mesh closely with the next section that will apply additional rules directly to construction contracts.

The descriptions are in short outline form to describe the principles in the most direct manner. They have been selected for inclusion here because of their common application to our industry. Discussions have been kept direct so they can be immediately understood and applied; discussion of theoretical reasonings are accordingly kept to a minimum.

Finally, it is important to be aware that discussions of "Contract" include the *entire* contract, including the plans, specifications, and all other *contract documents* as described in Section 2.4.2 of the Manual.

### 3.2.2 Reasonable Expectations

The first principle of contract law is to protect the reasonable expectations of the parties. The first step in doing this is to be able to confirm what those expectations are (or what they should have been)—to determine *whose* meaning is the right one.

In an effort to reduce subjectivity in the process, the courts have sought to develop an objective standard. The standard of a "reasonable expectation" has accordingly been described as the meaning that would be attached by normally intelligent people competent in their profession, with complete knowledge of all related facts.

#### Application rules

The following ideas will aid in the application of the principle as described above:

1. The ordinary meaning of language is given to words unless circumstances show that a different meaning is applicable.
2. Technical terms and works of art are given their technical meaning unless the context indicates a different meaning.

3. A writing is interpreted as a whole, and all writings forming a part of it are interpreted together.
4. All circumstances are taken into consideration.
5. *If the conduct of the parties defined a particular interpretation, that meaning is adopted.*
6. Specific terms are given greater weight than is general language.
7. Separately negotiated or added terms are given greater weight than standardized terms (boilerplate) or terms that are not specifically negotiated.

### 3.2.3 Ambiguities Resolved Against the Drafter

The law assumes that those drafting the contract (contract documents) will provide for their own interests and that they have reason to be aware of uncertainties. Not only should the contract drafters have had a clear idea of their intent, but they have had every opportunity to be sure that those intentions are clearly enough defined so as to allow reasonable, competent people to understand and accommodate them. The owner and design professionals typically may have had a year or more to be sure that the contract documents are clear and complete. It is unreasonable to expect a contractor to find flaws in the short time prior to bid.

It is important to be aware that this idea applies only to *adhesion contracts*—those that are produced by one party and presented generally in a "take-it-or-leave-it" fashion to another without any reasonable opportunity to negotiate. Public bids, for example, fit into this category. If on the other hand the contract itself is the product of extensive negotiation, it will have at least to a great degree been "drafted" by both parties. In a real sense, both parties will have been responsible for the ambiguity.

Even in that case, however, the negotiation may still have applied only to the Agreement portion of the contract, leaving the plans and specifications in the "adhesion" category.

### 3.2.4 Right to Choose the Interpretation

If anything is subject to more than one reasonable interpretation, the contractor has the right to choose the interpretation. The reasons are generally the same as those described for resolving ambiguities. If such a reasonable interpretation is available, there is no duty to continue to search for other "reasonable interpretations." All that is necessary is that your interpretation be reasonable; it does not even have to be the "most" reasonable.

### 3.2.5 Trade Custom

The Uniform Commercial Code defines "usage of trade" custom as

> ...any practice or method of dealing having such regularity of observance in a place, vocation, or trade as to justify an expectation that it will be observed with respect to the transaction in question. (U.C.C. Sec. 1-205(2))

A 2 × 4 is not 2″ × 4″, but 1 ½″ × 3 ½″. An 8″ CMU is 7 ⅝″. Customs always seem unusual at first, but through familiarity eventually make sense. If the designer referred to "a 2 × 4," you could rely on a dimension of 1 ½″ × 3 ½″. If, however, that designer referred to "2" × 4" blocking," be careful. Does the "Ambiguities" rule apply?

To establish trade custom as the definition of meaning, it must be shown that the custom is followed with *absolute regularity.* It is not sufficient that it is “usually” done in a certain manner; it must be a usage that is observed in virtually all cases in the area in which the contract is being performed.

Finally, trade custom cannot be relied on as an excuse if the requirement is clear and subject to one reasonable interpretation. When the roofing contractor says, “I never cut reglets,” read the specification. He may be cutting reglets on *this* job.

## 3.3 Key Principles of General Contracts

### 3.3.1 General

This section continues the principles of Section 3.2, applying the additional ideas directly to construction contracts. The descriptions are again in short outline form to describe the principles in the most direct manner, and have been selected for inclusion here because of their common application. Unique to our industry, these contract rights and requirements are very consistent throughout the majority of contemporary construction contracts. Discussions are direct so they can be immediately understood and applied; discussions of theoretical reasonings are again kept to a minimum.

Finally, as in Section 3.2, it is important to be aware that discussions of “Contract” include the *entire* contract—including the plans, specifications, and all other *contract documents* as described in Section 2.4.2 of the Manual.

### 3.3.2 “General Scope of Work”

The scope of work that must be completed in order to fulfill contractual obligations must be clearly and completely defined. Beyond that explicitly described the scope will be interpreted to include work that is *plainly and obviously* necessary for a complete installation. Strained interpretations that attempt to add work that is not specifically shown must be subjected to the most intense scrutiny. Consider the entire discussion of Section 3.2. From there, apply *“Intent” vs. “Indication”* (Section 3.3.3), *“Reasonable Review”* (Section 3.3.4), and *“Performance” and “Procedure” Specifications* (Section 3.3.6). Section 3.5 *Specification “Interpretation”* will also provide additional insight.

### 3.3.3 “Intent” vs. “Indication”

In a disagreement or dispute over scope of work, architects or engineers may try to apply what they *meant* to say if they cannot find anything in the documents that precisely indicates the item in question. Typically, some general conditions clause containing language to the effect that “It is the *intent*...to include all items necessary for proper execution and completion of the work,” may be relied upon by the owner.

The key to the applicability (or lack of) of design intent as a means to fill gaps left in the design goes to whether the work could be “reasonably inferable.” Design intent will apply only if the gap being bridged is so obvious that a professional contractor would not normally overlook it. Refer to *Trade Custom* (Section 3.2.5) and *“Reasonable Review”* (Section 3.3.4) for related discussion.

**Applicability test**

To assess the applicability of design "intent" to force a contractor to complete work not specifically indicated, answer the following questions:

1. Can the remaining work shown be completed without the extra work in question?
2. Is there more than one way to complete the extra work?
3. Is the extra work not usually encountered by the type of trade now being considered to construct it?

An affirmative answer to any of these questions will throw into question the applicability of "intent." For "intent" to apply, it will essentially need to satisfy the same test as that for trade practice (Section 3.2.5); that is, the work in question must be required with *absolute regularity* in all such cases.

### 3.3.4 "Reasonable Review"

Contractors are responsible for reviewing all bid and contract documents as they relate to their work. What can be seen, or *reasonably inferred,* gets priced. The contractor *is* responsible for the disclosure of *patent errors* (refer to Section 3.3.5), and the contractor is normally required to conduct a prebid site inspection.

While it is certainly desirable to be able to discover mistakes, inconsistencies, or other problems with the contract in time to minimize their effects on the project, it is not the contractor's *responsibility* to do so. The contractor is *not* required to perform a complete search of documents seeking out hidden flaws in the contract, making subsurface explorations, or undertaking any other extreme investigations. That should have been done by the owner and/or the design professionals prior to the bid.

### 3.3.5 Disclosure of Patent Errors

Beyond the responsibility for "Reasonable Review" as described in Section 3.3.4, the contractor may be held responsible for disclosing *patent errors.* These are mistakes so obvious or glaring (patent) that any competent contracting professional should have observed them through a reasonable review.

**Example:** A masonry contractor's experience will be that masons *always* incorporate flashing and ties in *every* cavity wall they ever built. If the details on *this* project, however, show no flashing or ties, the defect may be considered to be so obvious that the mason contractor should have at least raised the question.

It is interesting to note that the more "obvious" the error, the more in question the designer's competence becomes.

**Defense:** One argument for being paid extra for the work in question is simply that if the design mistake had been observed prior to bid, the cost for the additional work would have been added to the bid at that point. The greater risk, however, is that of consequential damage resulting from the work's being constructed with the error, and the necessity of removing incorrect work and rebuilding.

### 3.3.6 "Performance" and "Procedure" Specifications

A performance specification is one that described the ultimate *function* to be achieved, leaving the means and method up to the contractor.

**Example:** Provide foundation perimeter insulation having a k = .20 @ 75 degrees F, and a minimum compressive strength of 20 lb/ft.

A procedure specification describes the material to be used and its physical relationship to the surrounding construction. It should detail the qualities, properties, composition and assembly.

**Example:** Provide extruded closed-cell polystyrene board as manufactured by XYZ Corp., 2″ thick × 24″ wide.

**Application**

There is nothing inherently "wrong" with either type of specification. Performance specs, for example, may shift responsibility for the design onto the contractor, but they also increase the contractors options and control.

Typical contracts contain the largest percentage of procedure specifications. Product descriptions, however, often weave performance requirements into the same specification. This in effect duplicates the specification for the same product, adding the risk of indicating mutually exclusive requirements. In the case of such conflict, refer to *Ambiguities* (Section 3.2.3), and *Right to Choose the Interpretation* (Section 3.2.4) for related discussion.

Performance specifications may be used more often in descriptions of mechanical and electrical equipment, but they will even be used for things like concrete (minimum compressive strength of 3000 psi @ 28 days).

Procedure specifications must be correct in every respect in order to avoid problems. If it is properly followed, the risk of its design success will remain with the designer.

If a procedure *and* a performance specification are provided for the same item, the contractor is fulfilling its obligation by providing the exact material described. There is no further duty to confirm that the described material also meets the performance description—that was up to the designer in the first place.

### 3.3.7 Change Clause

The Change Clause authorizes the Owner to alter the work if the change in question falls within the general scope of the original agreement. If it does not, there is no inherent "right" of the owner to change the work.

**Basic elements**

Most Change Clauses will incorporate similar provisions. Typically, they include:

1. Adjustments to the contract may only be effected by a change order.
2. The change order must be in writing, signed by both parties.
3. The change order must specify adjustments to both the contract price *and* the net effect on the project time.
4. The change order will be for work that is within the scope of the original contract.

5. No changed work is to be performed without a properly executed change order (except in the case where the contractor must act in an emergency to prevent injury or property damage).

#### Caution

When asked to perform changed work without a finalized change order, resist. "Work now, work it out later" should have no place in your construction vocabulary. Use the change clause to display that you are not *allowed* to do the work unless the change is *signed*—and includes cost and *time.*

### 3.3.8 Pass-Through Clause

The Pass-Through Clause (sometimes referred to as the Conduit or Flow-Down Clause) incorporates into each subcontract by reference all the rights and responsibilities of the prime contract to the owner as they relate to the work of the respective subcontract.

**Example:** The general conditions may make "the contractor" responsible to provide "all scaffolding, hoisting equipment, etc., as may be necessary to perform the work."

The Pass-Through narrows this to mean the plumbing subcontractor provides all these items as necessary to complete the plumbing work, the plaster sub provides these items as necessary for the plaster work, and so on.

The authority of the Pass-Through comes from each individual specification section that references the general conditions, supplementary general conditions, etc., to be incorporated into the requirements of the respective technical specification section. It thereby brings all those requirements to bear directly on the work.

#### Application

Fortunately, the application of the Pass-Through is very straightforward. As you read any particular specification requirement, simply insert the name of the subtrade being considered wherever the word "contractor" appears; simultaneously insert the words "general contractor" wherever the work "owner" appears, and read the clause in that context.

**Example:** The general conditions may require that "the contractor will remove all rubbish ..." and "in the case of dispute, the owner may remove the rubbish and charge the cost to the contractor."

The Pass-Through will read this same clause as it relates to the millwork subcontractor as: "The millwork subcontractor will remove all (millwork) rubbish..." and "in the case of dispute the general contractor may remove the rubbish and charge the cost to the millwork subcontractor."

#### Exceptions

If the words "general contractor" or "prime contractor" have been used in the original specification, use caution. They may have been put there as a specific indication of just what it says—as an intentional modification of the Pass-Through.

**Example:** The *general* contractor may be specifically called upon to furnish, erect, maintain, and remove scaffolding for use by all trades in a certain area.

This may have been specified as such in order to avoid the extra time and expense if, for example, the mason were to erect and remove staging around the building only to be followed by the plaster subcontractor, who will then be followed by the painter—all repeating essentially the same work.

Additional common examples include the categories of temporary heat, dumpsters, and temporary utilities.

### 3.3.9 Dispute Clause

The Dispute Clause provides the specific procedure for resolution of serious problems. It may detail a progressive series of steps (such as appealing to higher authorities) or may simply describe the ultimate option (... arbitration ...).

Know what your Dispute Clause says, and always follow its instruction precisely. Don't make your problem worse by taking some action to resolve an issue that is not technically correct.

### 3.3.10 Authority (Formal/Constructive)

Descriptions of formal authority are spread throughout the general provisions of the contract. Beyond this, correct authorities are often confused because of familiarities with past relationships, or by the *constructive actions* of the parties.

Use extreme caution when determining where the formal authority lies for the situation at hand. The contract may "clearly" describe that the "commissioner" is the only one who can resolve the problem, but you may be able to show that those kinds of issues (changes of lesser cost, for example) were in fact "resolved" by someone else before, and that you've even been paid for it. You therefore had every right to rely on the *actual relationship* that was previously experienced that *constructively altered* the contract procedure.

Don't let others unnecessarily use boilerplate remarks to hide behind "authority." Use the actual job history to define the actual relationships.

### 3.3.11 Correlation of Contract Documents

Neither party to a contract is supposed to take advantage of any obvious error. Bad contracts do not live up to any technical or moral obligation to guide proper, responsible decisions. They cop out, trying to get the best of all worlds while covering up mistakes. They'll say something like "...in the case of a conflict, the more expensive detail will apply."

Better general contracts on the other hand will attempt to provide some mechanism to more definitely determine which kind of information takes precedence in the case of a conflict. (The Pass-Through Clause (see Section 3.3.8 of the Manual) ensures that the application will flow down to each respective "plans and Specs" subcontract.) This kind of treatment will also be found in most federal and many state construction contracts. If there is *no* language to guide you, Company Policy will follow the typical rules as stated below to resolve conflicts between the plans, specifications, and other contract provisions:

1. Amendments and addenda take precedence over the specifications.
2. The specifications take precedence over the plans.
3. Stated dimensions take precedence over scaled dimensions.
4. Large scale details take precedence over smaller scale ones.
5. Schedules take precedence over other data given on the plans.

In addition to these rules which hopefully will be clearly stated in the contract, correct legal consideration provides that specific terms, conditions, and requirements are given greater weight than is given general language designed to cover a multitude of circumstances (boilerplate). It is recognized that the specific language was designed with the intention to accommodate a unique circumstance. It is therefore reasonable to assume that it so considers the precise conditions. Supplementary General Conditions will therefore take precedence over the General Conditions. Refer to Section 3.3.12 *Specification "Boilerplate"* for related discussion.

### 3.3.12 Specification "Boilerplate"

Specification "boilerplate" is that language that is there supposedly to cover a multitude of common and "standard" conditions. There generally isn't any problem with these kinds of "general" provisions of the specifications that deal with *procedure* (general conditions, supplementary general conditions, working procedure). There are often problems with boilerplate that is placed there to deal with technical specification sections directly describing the work.

Because of its shotgun approach to work descriptions, it often conflicts with specific considerations necessary for particular situations, and just will not apply to the work or question at hand. In such cases, the language must be immediately recognized for what it is—excess baggage that causes inaccuracies and misunderstandings.

Never allow boilerplate to be used to justify a design oversight. It was placed there to cover a *standard* condition; don't let the general language be used to strain an interpretation for a piece of work that was simply missed by the designer.

### 3.3.13 Force Majeure

*Force Majeure* is a term given to delays that are the fault of neither party to the contract. They include those forces that are either unforeseeable, or otherwise beyond the control of either party, such as strikes, severe weather, and acts of God. These delays usually permit the contractor to file for and be granted an appropriate extension of contract time, but will not allow either party to be compensated for the time or expenses. The Owner correspondingly cannot seek actual or liquidated damages.

### 3.3.14 Impossibility and Impracticability

It's been said (courts have held) that there is no excuse for nonperformance due to misfortune, accident, or misadventure. This blanket approach, however, has many times been found to be too harsh. Exceptions to the rule have accordingly been explained in terms of constructive or implied conditions.

The doctrine of impossibility of performance is still evolving. One modern application has been determined as follows:

> A thing is impossible in legal contemplation when it is not practicable; and a thing is impracticable when it can only be done at an excessive and unreasonable cost.

The doctrine describes the gray area dealt with by courts in attempting to be responsive to practices in which the community's interest in having the contract strictly enforced is outweighed by the commercial senselessness of requiring performance.

#### Application

When the issue is raised, the condition of performance must be shown to be required by some changed circumstance in a process that involved the following steps:

1. A “contingency”—something unexpected—must have occurred.
2. The risk of the unexpected occurrence must not have been allocated to either party by agreement or “custom.”
3. The occurrence of the contingency must have rendered the performance commercially impractical.

Note that although extreme impracticability (financial hardship) provides the basis for argument, mere “unanticipated difficulty or expense” not amounting to impracticability is not within the scope of this application.

As straightforward as the above ideas are, their extreme difficulty in application lies in the agreement between the parties of items 2 and 3.

**Example:** If, for example, the concrete slab varies within approximately ½″ in 10′, is it “reasonable” for the architect to insist that it be removed and replaced because the specification stated that concrete slabs shall “be level within ¼″ in 10′ in either direction?”

The cost for such replacement is high. It may actually be reasonable if the contract contemplated subsequent installation of scientific equipment or the installation of some following construction that required those tolerances. It may, however, not be so “reasonable” when it is a shop floor that is only going to get beat up with tow motors.

### 3.3.15 Termination

The remarks here apply to agreements between the contractor and owner agreements and those between the contractor and subcontractor.

For any termination to be appropriate, performance must be so poor as to be considered a “material” breach (as opposed to a “minor” or “immaterial”) breach. The victim of a material breach may regard the contract as having been abandoned or terminated. The victim of a minor breach may sue for damages, but must continue its performance under the contract.

Because of the enormous significance of any move to terminate, the determination of what constitutes a material breach versus a minor one becomes *the* question.

#### Termination clauses

In an attempt to avoid problems associated with the distinction between “material” and “minor” breaches of contract, some contracts include a provision that allows for termination for “convenience,” or specifically for those cases that will not be considered “material.” This does not however, automatically mean that the terminated party will not be compensated for the damage sustained by such a termination.

Other contracts more directly recognize the crucial significance of determining whether a breach is material or minor by specifically listing categories of actions by which both parties to the contract essentially agree to in advance are significant enough to warrant termination. Such a list might include:

1. Serious and/or repeated violations of federal, state, or local laws and ordinances

2. Consistent failure to provide labor, material, equipment, and supervision to maintain adequate rates of progress
3. Significant or repeated failure to provide and install work in conformance with the contract documents
4. Consistent interference with the work of others by failing to properly coordinate the work, both in construction details and in sequence—manner of completion
5. Failure to pay subcontractors or suppliers for materials and labor acceptably supplied to the project
6. Filing for bankruptcy, making an assignment for the benefit of creditors, or committing other acts of bankruptcy
7. Subletting significant portions of a subcontract without approval to unqualified vendors

A termination clause does not *require* a party to terminate the contract if there is a clear right to do so; it is merely an option. The party considering such a termination could instead allow the offending party to complete the contract, and sue for damages related directly to the potential cause for termination:

**Example:** Rather than terminate a subcontractor for repeatedly failing to arrive at the site to install a certain portion of the work, or because of its having installed substandard work, the general contractor may arrange to have that portion of the work completed by other means, backcharge the offending party for all direct costs incurred, and sue for damages directly related to the poor work and resulting delay.

Termination is *very* serious business. It should only be seriously considered in the most extreme and chronic cases of contract breach. If termination must be pursued, senior management must be made aware of all details of the breach and proposed action and *must sanction the move.*

The termination itself will eliminate all chances of settling the dispute between the parties. Potential damage to the terminated party is so significant that you can certainly expect a fight to prove that the termination was wrongful. The rule, therefore, is this: it is not enough to be 100% correct in your position; you must be at *least* 300% correct.

### 3.3.16 Notice

In a modern general contract, there will be literally dozens of references to some kind of "notice" requirement as related to almost anything. Most critical will be those relating to changes and claims, but they can apply to many other items.

It is not necessarily the date of the first formal correspondence that qualifies as the effective notification date. Notification has been achieved if there has been an understanding in the mind of the recipient. The purpose of "written" notification is simply to remove doubt as to when this had been achieved. Notification, therefore, can sometimes be said to have occurred in many types of communications, including:

- Telephone or other conversations
- Meetings

- Other letters
- Shop Drawings

It is not a good feeling to have lost a legitimate issue on technical notification grounds; don't put yourself or the Company into that position. When in doubt, get *some* kind of notification confirmed.

Such notification for a specific kind of issue (change order, for example) will be required to be in writing, and delivered within a precise number of days of the issue's "occurrence." It is important to realize, however, that it can be argued that the operative date beginning the notification period will be from the point at which it was first realized that there will definitely be an effect on the contract time and/or cost—not the point at which it was simply realized that an event occurred.

Be aware of your notice requirements and comply with them in all cases. In so doing:

1. *Always establish the earliest possible legitimate date.* If the issue was discussed at a job meeting or in a phone conversation, refer back to it in the first part of the notification. "Confirming our conversation of (date)," will be enough.
2. *Notify everyone who might possibly require it.* Do this either directly or by copy.
3. *Specifically name the individual(s) involved in the prior notification.* "Confirming my conversation with your office," is too weak. Name names.
4. *Get into the habit of confirming all potentially significant discussions immediately.* Any piece of information that has any potential to affect the project should be recorded in a manner that corresponds to the level of potential effect. Even if a seemingly insignificant item begins as a plain note in a file, it can form the basis of a later formal notification if one becomes necessary.
5. *Be concise, but be clear.* Refer to Section 2.3.2, *Rules of Effective Correspondence Project,* and apply them in all your communications.

## 3.3.17 Proprietary Specifications

#### General

A proprietary specification is one that limits competition. On private contracts, owners have the right to specify exactly what they want if they're willing to pay for it. Although there are still formal and informal procedures to consider "equals" and "substitutions" (see Section 3.7), there is no obligation on the part of the owner to provide for fair competition among competing products.

On public contracts, however, the owner *does* have such a responsibility. Specifications are supposed to be written with the clear intention to encourage competition, and allow as many vendors as reasonable possible access to the project's "market." To this end, each product description should name at least three "acceptable" manufacturers and/or add the words "or equal" to the list of named sources. In the case where this language treatment has not been complied with, public policy, state statute, or federal regulations may provide the needed basis to allow competition.

#### Application

Although the specification may appear on the surface to comply with the requirements by naming alternative sources and by using the words "or equal," the difficulty may next lie in the fact that except in the most simple product descriptions (like a steel stud) it is rare to find two products that are made precisely the same, or have the same list of technical specifications.

The problem is compounded with the complexity of those product descriptions. A Douglas fir stud is a Douglas fir stud, but carpets will have two dozen or so technical criteria, with no two carpets having the same list. The problem may be even worse in the case, for example, of mechanical equipment.

Read Section 3.7 on "equals." If your proposed product is being rejected because it does not precisely match up to a long list of technical items, the specification may be considered to be unnecessarily restrictive so as to illegally limit competition, or the problem may simply be the designer's placement of an inappropriately large amount of weight on some technical criteria that does not go to the essence of the product needed. Consider again the definition of *equal*: "The recognized equivalent in substance, form, and function...."

## 3.4 General Contract Bonds

### 3.4.1 General

A bond is the guarantee of one party for the performance of another. Construction bonding is essentially a three-party contract between the principal (the contractor), the obligee (the owner), and the surety.

In a real sense, the first purpose of a bond is to identify the actual ability of a contractor to get one. This ability tends to separate unqualified contractors out of the process. Before any surety guarantees the performance of any general contractor, it will have conducted the most intrusive and detailed investigations of the contractor's financial strength to carry the type of work contemplated, and of its management ability to deal with all factors of production. The company principals will personally have had to demonstrate absolute commitment to the surety for its actual performance by providing large amounts of financial security to back its promises of performance and competence. Nowhere else is it more true that a bonded contractor has "put his money where his mouth is."

Finally, it is important to note that every bonded contractor will have a profound, sober understanding that if it ever needs to rely on a surety to cover performance, it will be the one and *only* time. The contractor not only risks the security provided for the bond, but will in all likelihood be unable to get another bond. Except in the most unusual circumstances, the contractor will effectively be on the way out of business—or at least out of its ability to bid bonded work.

There are many types of bonds. As related to general construction contracts, the prevalent ones are bid, payment, and performance bonds.

### 3.4.2 Bid Bonds

A bid bond is an assurance to the owner that if selected, the contractor will actually proceed with the contract at the bid price. If the contractor does not, the bid bond becomes payable to the owner as compensation for damages sustained.

The measure of damages is usually a penal amount specified in the bond, or the difference between the amount of the bid of the defaulting contractor and the next highest bid (if the difference is less than the penal amount), or in the absence of a higher bidder, the price at which the owner will ultimately be forced to contract for. The most common penal value is 5 percent of the amount of the bid. Where such a maximum penalty is specified, that amount is the limit of both the contractor's and surety's liability, even if the owner has sustained damages beyond such a limit.

**Application issues**

1. *The bid bond may not stipulate that the surety providing the bond will also provide subsequent payment and performance bonds.* Such language protects the surety against having to provide such bonds for questionably bid contracts that would have been awarded. For this reason, many bid procedures also require a surety letter of intent to provide payment and performance bonds if the contractor is awarded the project.
2. *Bid mistakes may be justification for not being held responsible for a bad bid.* Refusal of a successful bidder to go ahead with the contract is usually because of a significant error in the bid. If the contractor can offer "clear and convincing" proof of such an error, there is a chance that both the contractor and surety will be relieved of the liability. Rules of such a mistake are tough, and include the ideas that the mistake:

   - Must be one of fact, and not one of law or judgment
   - Must be of such grave consequences that enforcement of the contract would be unconscionable
   - Must relate to a material feature of the contract
   - Must not have come about through the contractor's culpable negligence
   - Must not prejudice the owner except to loss of bargained-for performance

   It should be clear that if the contractor's mistake was this bad, the surety will not jump at the chance to guarantee the contractor's next bid.
3. *A material change to the contract will void the principal's obligation to the bid.* If a material change in the terms and conditions of the contract that could not be reasonably anticipated when the bid was submitted occurred after the bid, the principal may be entitled to withdraw its bid.

   Examples of material changes may include:

   - Delay in awarding the contract beyond the time specified in the invitation to bid
   - Changes in the method of construction from those specified in the bid documents
   - Changes in the materials specified
   - Massive "corrections" of documents that otherwise would have resulted in significant change to the bid price

### 3.4.3 Performance Bonds

The performance bond protects the owner from the contractor's failure to complete the contract in accordance with the contract documents by indicating that a financially responsible party stands behind the contractor to the limit of the penal amount of the bond. Most performance bonds will accordingly require such a guarantee for 100 percent of the amount of the contract price. The entity protected is usually only the named obligee (the owner), with such protection not usually extending to third parties.

The Miller Act of 1969 requires performance bonds on all government projects. For other projects of significant size, it is much more than just a good idea for the owner.

The performance bond, however:

1. *Does not provide any guarantee that the contract work will be completed as specified for the contract price.*
2. *Limits the surety's liability to a specific dollar amount—the cost to complete the work.* This will not normally include consequential damages.
3. *Can be discharged if the owner permits a cardinal change in the contract.* A cardinal change is one that fundamentally alters the scope of contract performance.
4. *May be discharged if the owner violates contract terms that are prejudicial to the surety.* Failing to provide builder's risk insurance, for example, may be such a violation.

### 3.4.4 Payment Bonds

Labor and Material Payment Bonds protect those who have supplied material and labor to a project, first because there may be no lien rights against public properties. The bonds also protect owners from liens or other claims made against the property on nonpublic projects after completion of the work, and after final payment has been made to the contractor.

The Miller Act provides direct protection by way of Payment Bonds for *those who have a direct contractual relationship with the prime contractor or with a subcontractor.* Accordingly, those who furnish material or services to sub-sub's materialmen or third-tier subcontractors would not be covered by the Act.

The bonded contractor must therefore police all payments made through the chain of subcontractors and primary suppliers to be certain that all payments have in fact been made to all intended suppliers and materialmen. If such payments are misapplied, the bonded contractor may be required to pay twice for the same work.

Typically covered items include:

1. Materials incorporated into the work, delivered to the jobsite, or furnished pursuant to the contract documents even if not delivered to the site
2. Labor performed at the jobsite, and labor performed in fabricating materials off site pursuant to the contract
3. Freight and transportation costs
4. Equipment rental and repair costs
5. Fuel and maintenance items
6. Insurance premiums
7. Unpaid withholding taxes (on federal projects)
8. Union pension and welfare benefits
9. Some categories of legal interest and attorney's fees

Items usually not covered include:

1. Bank loans (even if the loans were used to pay project costs)
2. Claims of liability or damage arising out of performance of the contract
3. Withholding taxes on state or private contracts

Note that in the determination of items to be covered, it has been upheld in many states that a supplier only need to "reasonably believe" that its products were shipped to your jobsite. If a subcontractor, for example, picked up materials and noted your job as their destination on the receipt, you may have a problem.

Likewise, some subcontractors may arrange to have material shipped and even delivered to your job because the job is nontaxable, then take the materials to another project. If you allow such a thing to go undetected, you may be ultimately held responsible if there is a payment problem involving those materials.

## 3.5 Specification "Interpretation"

### 3.5.1 General

The general conditions is likely to contain some language to the effect that "The architect will interpret the requirements of the plans and specifications...," and "The architect shall interpret the requirements of change orders and shall decide all other questions in connection with the work."

This is a common trap. The same specification, for example, is likely to go on to include something like "The architect shall have no authority to approve or order changes in the work which require an extension of time or change in the contract price."

On public contracts, the language will be modified to place authority (and *responsibility*) for interpretation more directly on the owner representatives, thus reducing the architect to an advisory capacity.

### 3.5.2 Application #1: Right and Duty

The question of right or duty to interpret the documents is closely related to the discussions on the *Change Clause* (Section 3.3.7), the *Dispute Clause* (3.3.9), and *Authority* (Section 3.3.10). It is interesting to note that the architect and engineers will usually be anxious to apply their "right" in cases where the disagreement is over a "missing" item (see Trade Custom (Section 3.2.5) and Ambiguities (Section 3.2.3). In cases of *conflict* over redundant or missing information between or within specification sections, however, these same individuals will energetically avoid their *responsibility* to interpret. Correspondingly, the owner representative will often be content to allow the process to play itself out between the construction and design forces until finally forced into the decision process.

**Example:** If two subcontractors both take exception to performing a certain scope of work, the general contractor may unintentionally and unnecessarily assume responsibility for the decision by making it directly. At the job meeting, for example, you will be told something like "... that's between you and your subs; it's in the spec." Nonsense. It is *their* specification, and it says *they* will interpret it. Point out the appropriate general conditions section that requires that the architect (or owner) interpret the documents, and that you are accordingly entitled to be directed *precisely* where the work in question is specified—what specification section, what plan details, etc., give the authority. From that indication, you can give a contractually responsible decision to the respective subcontractors involved. If it is a good decision, you will have complied with the documents, maintained credibility, and kept responsibilities where they belong. If it is a bad decision, you will preserve the ability to proceed with the work by a contractually proper owner directive to the subcontractor, while keeping the ultimate *liability* for the decision where it belongs—with the owner.

Refer to Section 3.1.2 *G.C. as "Conduit"* for important related discussion.

### 3.5.3 Application #2: Appeal to Higher Authority

In many cases either at the job meeting or correspondence level, your position on a matter will be rejected by an "authority" that may or may not be appropriate at that level. Such a rejection may also be accompanied by either poor grounds or no grounds at all. In this situation, begin by confirming who should have been making the decision at this level in the first place. (You should have already considered this in the original correspondence; see Section 2.3.2.) Even if there is a clear contract statement to the effect that "the architect shall interpret the documents," move next to the dispute resolution structure of the contract. Consider the *Change Clause* (see Section 3.3.7), the *Dispute Clause* (Section 3.3.9), and any other clause that qualifies the architect's "final" decision. Even where "the owner" makes the final interpretation (on public contracts) the first "decision round" is normally made at the jobsite level by the Clerk of the Works, Construction Supervisor, or a person in a similar position. But *read* the spec. It may actually say something to the effect that "decisions of the commissioner will be final." If so, you have every right to *get* to the commissioner if lower level decisions do not satisfy you.

**Example #1 (Public Projects):** The architect rejects your argument at a job meeting, and "decides" that you are responsible to provide something not shown. Stay cool. Turn to the Owner representative, demonstrate his/her responsibility to "interpret the documents," *clearly* summarize your position (with all its backup) and the architect's position, and require the *owner's* interpretation.

**Example #2 (Municipal, or Other Public Funded Projects):** Many project owner structures incorporate the final authority of a building committee. In these cases, that will be the point of final appeal. If so, proceed to request the opportunity to present the matter directly to them.

In some cases, the demonstration of that kind of determination may be sufficient to bring a relatively small matter back under control. The project representatives, for example, may have an interest in avoiding burdening the entire committee with problem "details." Who knows, they may just like to avoid staying up all night at the meeting.

The dilemma is that you will (as in the other case) be asking them to find in your favor to their own disadvantage. Although this will always be a problem, the good news is that most committees take their responsibilities very seriously. Further, in serious cases, they will have the ability to refer to an attorney for direct advice. If you are properly and clearly supported, have done your homework, and have made a clear presentation, you stand a good chance of turning their heads.

### 3.5.4 Sample Letter to the Owner Regarding Specification Interpretation *(page 3.21)*

The Sample Letter to the Owner Regarding Specification Interpretation is an example that incorporates the ideas of this section. Specifically, it:

1. Summarizes and *accurately* represents the statements and references by different subcontractors
2. Identifies the contract reference to those directly responsible to "interpret the documents" to make the *decision*
3. Puts some weight behind the responsibility to interpret by directly notifying of additional problems resulting from failure of the owner to properly act

## 3.6 Shop Drawing "Approval"

### 3.6.1 General

The concept of shop drawings, or "approval submittals," is constantly being tried—by design professionals and subcontractors and materialmen—both sides of the coin. Motivations are rooted in everything from simply trying to avoid the time and effort necessary to do the job right to direct and strained efforts to avoid responsibility and liability.

On the design side, chronic problems exist such as:

1. The use of "creative" words and language on shop drawing stamps
2. Actual failure of the design professional to check for compliance with design intent
3. Direct attempts simply to avoid having to spend the time necessary for proper review and other action that the construction force may be entitled to (particularly on equals and substitutions)
4. Failure to take action in a reasonable amount of time
5. Taking inappropriate full or partial rejection positions
6. Refusing to accept the consequences of his or her "approval," regardless of what the stamp might say

On the side of those who prepare, submit, or pass-through submissions of shop drawings, problems include:

1. Failure to comply with contractual requirements regarding the manner, scale, amount, and type of information presented in the submittal
2. Failure to submit such documents in time to avoid delay in the work
3. Failure to coordinate the work with contiguous work
4. Attempts to avoid responsibility for properly accommodating field dimensions or other field conditions
5. Failure to provide sufficient number of copies for proper distribution
6. Failure to properly review third-tier submittals for compliance with the contract
7. Failure to properly review third-tier submittals for compliance with the primary subcontract
8. Failure to clarify responsibilities between primary vendors or subcontractors and their subs or suppliers
9. Inappropriate treatment of "equals" and "substitutions"
10. Failure to properly distribute information to all those who might need it—in time to avoid conflict

## 3.5.4
## Sample Letter to the Owner Regarding Specification Interpretation

**Letterhead**

Date)

To: (Owner)

RE: (Project)
(Company Project #)

SUBJ: Welded Portion of Masonry Anchors

Mr. (Ms.) ( ):

Section 04200 3.4 requires the masonry contractor to provide the welded portion of the masonry anchors attached to steel columns. Detail 4/S3 indicates the same work to be performed by the structural steel subcontractor. Both contractors have refused to perform the work as part of their contract.

General Conditions Article 20 directs that the owner interpret the requirements of the contract. Accordingly, please advise of the party responsible for the subject work, indicating the applicable specification(s) and the responsible contractor by name.

Your complete written response is required by (date) in order to avoid (additional) interference.

Very truly yours,

COMPANY

Project Manager

cc: Everyone named in the letter
Architect
File: Masonry
Str. Steel
CF

Actual procedures for securing, reviewing, and processing all approval submittals are treated in great detail in Section 4 *Project Engineering.* This section will focus on contractual responsibilities and liabilities in the submittal process, and is closely related to the discussion of *"Equals" and "Substitutions"* of Section 3.7.

### 3.6.2 "Approval" Abuse

If architects and engineers wish to hold themselves as "professionals," the public, and contractors, have the right to expect that they behave as "professionals" and to be held to certain standards of conduct.

Professional liability insurance premiums and overall litigiousness of the industry have caused a preoccupation among architects and engineers with ways to reduce their legal exposure and liability.

The most common first-line approach architects and engineers use to this end has been to change the "approval" statements and to add strained, exculpatory language to their shop drawing approval stamps:

1. The word "approved" has been totally eliminated from their vocabulary. Words like "No Exceptions Taken," "Examined," "Reviewed," and (my favorite) "Not Rejected" are used in attempts to somehow reduce their liability.
2. Paragraphs describing what they are and are not doing make the stamp larger than the paper that was stamped.

In most instances, these efforts have produced no beneficial effects, and in many instances, they have had decidedly adverse effects:

1. Courts have clarified that the use of words other than "approved" does *not* bar a liability. They acknowledge that:
   *a.* Only the person with the "big picture" could carry the responsibility for design safety and integrity.
   *b.* The contractor relies on the architect's written authorization to proceed with the work at all. Whatever the actual word may be, it is considered as an express statement authorizing the work to proceed as described in the submittal. It cannot be construed to mean "it's OK to fabricate and install the work this way for now, but I may reject it later."
2. The inclusion of detailed language may actually be an express statement by the designer that he or she is actually *not* performing the reviews or taking the action that is required by the owner/architect agreement, the owner contractor agreement, or even the general conditions of the contract. He or she may therefore actually be setting themselves up for a case of gross negligence.

### 3.6.3 General Contractor Liability

Under prevailing standard general contracts, both the general contractor and the design professionals share a responsibility for review and "approval" of submittals. It has been arranged in this way in an effort to increase the likelihood that shop drawing errors will be detected and corrected.

Contractors' review/approval responsibilities principally are to:

1. Check for conformance to the contract documents
2. Highlight deviations from those documents, either because of necessity (conflict, impossibility, etc.) or other reason
3. Coordinate the work with contiguous work

Unless there is extremely unusual language in the general contract, it is *not* the contractors' *responsibility* to find mistakes or otherwise correct the contract documents themselves.

In recent years, there has been at least some discussion of eliminating the designer's approval role altogether as a further effort to reduce designer liability. This route has not gained much support, however, principally because of the potentially high risk to public safety. But the point is, contractors must be aware on a project-by-project basis of the precise responsibilities that are being assumed for the particular contract.

### 3.6.4 Appropriate Contractor Action

The specific shop drawing review procedures for both subcontractor and in-house submittals are detailed in Section 4 *Project Engineering*. The action described here is to control any abuses attempted by design professionals to avoid approval liability.

Discussion of related actions regarding *Equals and Substitutions* is incorporated in Section 3.7.

When confronted with any action on your submittals that is other than "approved," be aware that:

1. The owner/architect agreement is almost certain to contain language requiring that the architect *approve* submittals. There won't be anything requiring the architect to "not reject."
2. The owner/contractor agreement will instruct the contractor that "the architect will review and *approve*" your shop drawings. You have an absolute right to rely on such statements.

Take the following steps to correct, or at least clarify, the meaning of language and reliance of the construction force on it to allow construction to proceed:

1. Advise the Owner and design professionals of their responsibility to *approve*, per the owner/contractor agreement.
2. The designers will likely refuse. They may even admit openly that it is their professional liability insurance company that is preventing them from using the word "approved."
3. Point out the facts of the owner/architect agreement and the owner/contractor agreement as described above. Unfortunately, this will still in all likelihood not induce the design professionals to change.
4. Point out the apparent effort to avoid approval responsibility as described in the contract, and the practical problem of not being able to release anything for construction until the owner clarifies just what "words" will so allow the work to proceed.

5. If the Owner and design professionals agree to change the language, don't wait for it. Send the *Sample Letter #1 to the Owner Clarifying Shop Drawing "Approval"* of Section 3.6.5, thereby:

   - Confirming the decision
   - Making the action retroactive, in order to avoid having to resubmit everything that has been acted on to that point

6. Unfortunately, it is more likely that resistance to changing the stamps will be great. When met with refusal to alter the language, be firm about the purpose and the effect on the project. Take all steps necessary, including the immediate involvement of Company senior management, to confirm the owner's proper operative definition of whatever words will be used. When such "clarification" is secured, confirm it back to the owner using the *Sample Letter #1 to the Owner Clarifying Shop Drawing "Approval"* of Section 3.6.5, thereby:

   - Confirming the operative definition of their word of choice as actually meaning "approved," whether they like it or not
   - Demonstrating that the entire construction force is relying on that direction as authorization to proceed with the work as described in the respective submittal

### 3.6.5 Sample Letter #1 to the Owner Clarifying Shop Drawing "Approval" *(page 3.25)*

The Sample Letter #1 to the Owner Clarifying Shop Drawing "Approval" accommodates the ideas of Section 3.6.4 for a condition in which the owner agrees to alter the language to be used on the shop drawing stamps. Specifically it:

1. Confirms the Owner decision to require the design professionals to correct the language on the shop drawing approval stamps.
2. Defines the words that had been used so far as meaning "approved," thereby eliminating the necessity of resubmitting previous submittals with the otherwise inappropriate language on the stamps.

### 3.6.6 Sample Letter #2 to the Owner Clarifying Shop Drawing "Approval" *(page 3.26)*

The Sample Letter #2 to the Owner Clarifying Shop Drawing "Approval" follows through on the suggestions on Section 3.6.4 in the face of a condition where the owner "agrees as a practical matter" with the purpose of the words on the shop drawing stamp, but does not ultimately force the design professional to change it. Specifically, it:

1. Defines those words that will be construed by the construction force to actually *mean* "approved," regardless of what the word actually is, along with whatever qualifications that may have been part of the discussion
2. Notifies the Owner that the construction force is relying on those words specifically as the owner direction to proceed with the work as it is described in the submittal

## 3.6.5
## Sample Letter #1 to the Owner Clarifying Shop Drawing "Approval"

**Letterhead**

(Date)

To: (Owner)

RE: (Project)
(Company Project #)

SUBJ: Shop Drawing Approval

Mr. (Ms.) ( ):

This letter is to confirm our discussion of (date):

1. The design professionals have been using shop drawing approval stamps with the words "No Exceptions Taken" in lieu of "Approved," and "Note Markings" in lieu of "Approved as Noted;" contrary to the requirements of General Conditions Article ( ).

2. You confirmed today that the stamps will be corrected to reflect the proper use of the word "Approved" and "Approved as Noted;" the other language will be dropped.

3. In order to avoid the necessity of having to resubmit and redistribute all submittals that bear the words "No Exceptions Taken" or "Note Markings," you confirmed that on those submittals the words "No Exceptions Taken" are construed to mean "Approved," and the words "Note Markings" are construed to mean "Approved as Noted."

4. In order to allow the work of those previous submittals to proceed without further interruptions, please confirm letter by signing it in the space provided and returning it to my attention.

Agreed to By: ______________________________ Date: ____________

Very truly yours,

COMPANY

Project Manager

cc: Architect
File: Submittal Approval
CF

## 3.6.6
## Sample Letter #2 to the Owner Clarifying Shop Drawing "Approval"

| Letterhead |
|---|

(Date)

To: (Owner)

RE: (Project)
(Company Project #)

SUBJ: Shop Drawing Approval

Mr. (Ms.) ( ):

This letter is to confirm our discussion of (date):

1. The design professionals have been using shop drawing approval stamps with the words "No Exceptions Taken" in lieu of "Approved," and "Note Markings" in lieu of "Approved as Noted;" contrary to the requirements of General Conditions Article ( ). General Conditions Article ( ), however, states that the architect will approve shop drawings. We therefore require that the design professionals correct their stamps to properly reflect such approvals. Note that the owner is as much entitled to this security from the designer as the construction force is, by virtue of its owner/architect Agreement.

2. Although you agree that the constructive purpose of the architect's action on such submittals is to release the item for use in the project, you have refused to require the design professionals to actually correct their stamps' language to meet the requirements of the contract.

3. Accordingly, in order to avoid delay to the project, the words "No Exceptions Taken" will be construed per your direction specifically to mean "Approved, and "Note Markings," will be construed to specifically mean "Approved as Noted;" all strictly within the context of the owner thereby expressly directing the work to proceed as indicated in the respective submittals. These operative definitions must be relied upon in order to allow any work to proceed. Please confirm your agreement by signing this letter in the space provided and returning it to my attention.

Agreed to By: ______________________________ Date: __________

Very truly yours,

COMPANY

Project Manager

cc: Architect
File: Submittal Approval, CF

# 3.7 Equals and Substitutions

## 3.7.1 General

The concepts of "equals" and "substitutions" are very often misapplied by owners and designers. Unless corrected, they almost always automatically place "equal" products and submissions immediately into the "substitution" category in order to force the contractor either to forget using the product (reject the submission) or to secure some kind of credit or other "reimbursement," if they decide that they can actually live with the product.

Reasons for rejection range from an honest objection to the product's being proposed for some otherwise understandable problem to the idea that the designer's office simply doesn't want to spend time reviewing any products that it had not specified.

## 3.7.2 Typical Contract Treatment

Many general contracts contain language, either in the general conditions or in the bid documents themselves, requiring that *substitutions* be submitted within a certain number of days after contract award, or after the bid, or even with the bid. This will be the first line of attack on the submittal for any product that was not named in the specification.

Because of the realities of project buy-out logistics, it is usually improbable that all bid items will be bought within those usually very short time periods and that it will be practically possible to secure all project submittals within such aggressive time frames. The other hurdle will be the treatment of the submittal itself, whether or not it satisfies any "substitution" time requirement. The designer is simply going to want what was specified, for good or bad reasons. Be prepared, therefore, for an uphill battle. Consider Section 3.3.17 *Proprietary Specifications* for related discussion.

Better contracts will go on to define exactly what products will be considered "equal," and what will be considered a "substitution." Know in every case precisely what the contract language is, and deal with it directly. In the absence of such clear definitions, the following are the operative ideals:

**Equal**

The recognized equivalent in substance, form, and function, considering quality, workmanship, economy of operation, durability, and suitability for the purpose intended, and not constituting a change in the work.

**Example:** If a 2 × 6 16-ga. structural steel stud manufactured by ABC Corp. is specified, a 2 × 6 16-ga. structural steel stud manufactured by XYZ Corp. should be considered an *equal.*

**Substitution**

A replacement for the specified material, device, or equipment which is sufficiently different in substance and function *to be considered a change in the work.*

**Example:** If a 2 × 6 16 ga. structural steel stud manufactured by ABC Corp. is specified, a 2 × 6 Douglas fir stud should be considered a substitution.

## 3.7.3 Application

Refer to Section 3.1.2 *GC as "Conduit"* for discussion related to principles of action on the part of the general contractor taken on behalf of, or otherwise directly at, subcontractors and suppliers.

Refer also to any particular subcontract that is involved with a specific "equal" or "substitution" issue to confirm that valuable time and energy are not being wasted on subcontractor submissions of which that subcontractor is not entitled to by the terms of its subcontract.

Submissions for product substitutions and equals from subcontractors and suppliers will be accepted by the Company for submission to the project designers only after it is confirmed that they have first met all relevant requirements of the individual subcontract or purchase order. If they have not, it is important to understand that processing such a submittal to the designers will likely expose the Company to otherwise unnecessary coordination, late deliveries, and even possible product liability. These considerations must, however be weighed against the right of the design professional or owner to interpret the documents. You may after all be forced to pass through the submittal for their decision anyway. If this turns out to be the case, get it out and back as quickly as possible.

After specifically considering these references and concerns, proceed as described below for those submissions so confirmed to be appropriate.

#### Substitutions

Double check that the submission is in fact a substitution. Don't assume too easily. Is it really a *change* in the work? Legitimate reasons to consider a real substitution include:

1. The material described in the specification is no longer available.
2. It can be reasonably shown that the material described in the specifications is inadequate, inappropriate, or will not otherwise serve its intended purpose for whatever reason. Remember—substitutions do *not* necessarily mean a credit change order.
3. Material deliveries or other problems related to the specified material or equipment or its supplier will cause unacceptable delivery or erection schedules or will otherwise affect contiguous construction.
4. There is significant price and/or construction advantage with a proposed substitution to warrant the effort necessary for its approval, considering the strong possibility of having to offer an acceptable credit.

Upon such reconfirmation, comply explicitly with any/all stated submission requirements for substitutions, considering:

- Time of submission
- Number of copies
- Form, manner, and content of all submittal information
- Requirements for comparative information on "specified" items

For both equals and substitutions, there is usually some requirement to include a complete submission of the originally specified item, in order to allow the reviewer to conduct a side-by-side comparison of all product features and information. This is usually the third place where a subcontractor's submission will be deficient (first: it will be late; second: there will only be one copy).

If such requirements are so explicitly stated, there is just no excuse for an improper submission. Don't give anyone such a simple excuse for rejecting the submittal.

**Exception:** Submissions requiring preparation of detailed, individually drafted drawings, or other type of submission that will require elaborate, expensive preparation will just not be available from those (originally specified) vendors—at least not without considerable expense. In this case, point out the practical problem, and submit on the product itself, complying with all submissions requirements.

After all these requirements have been met, follow the procedures described in Section 4.10 *Submittal Requirements and Procedures* to expedite the actual submission.

#### Equals

The procedure for treatment of submittals is similar to that for substitutions. The differences lie in the need or right of the GC or sub-vendor to make such a change, and the final treatment of the submittal by the contract. True "equals" stand a much stronger chance of seeing the light of day:

1. Review, as with substitutions, the respective subcontract or purchase order to confirm that there's a need for any action other than insisting that the sub-vendor provide specified material. Upon such reconfirmation, comply explicitly with any/all stated submission requirements for substitutions, considering time, quantities, and form.
2. Process the submittal in accordance with the procedures described in Section 4.10 *Submittal Requirements and Procedures.*
3. Consider if the specification can be considered proprietary (refer to Section 3.3.17). If so, consider calling attention to this idea in the submission transmittal.
4. Be prepared to have your submission treated as a substitution, with the result that the first designer reaction to it will be flat rejection on inappropriate grounds. Consider ways to head off this time-wasting posturing.

### 3.7.4 Perspective

Throughout the substitution or equal process, it is important that the general contractor maintain appropriate perspective. Remember the *GC as "Conduit"* idea of Section 3.1.2. In one real sense, the GC is the custodian of each subcontractor's rights relative to the owner's general contract, and as such, must take all actions regarding its submissions as responsibly as possible. On the other hand, that subcontractor is signed to the same responsibilities as you are (refer to the Pass-Through Clause of Section 3.3.8), and the GC in the final analysis has no "right" to interpret the documents (refer to Section 3.5).

When an owner's (designer's) rejection of an equal or substitution is received, it is important that you transmit that action clearly, completely, and immediately to the vendor affected, also transferring the designer's direct order to provide that which is specified. If you maintain this perspective, you will:

1. Keep the sub-vendor responsible for the product, for its effect on contiguous construction, and for its delay that it subjected the project to—if the designer is right.
2. Have kept all liabilities for extra cost and extra time that may ultimately be determined to have not been the fault of the sub-vendor or the owner—if the designer is wrong.

## 3.8 Responsibility to "Coordinate": Use and Abuse

### 3.8.1 General

As a general contractor or construction manager, your responsibility is to coordinate the *work*. It is not to invent information and not to coordinate the design. When a problem surfaces, it has become common practice to criticize the contractor for failure to coordinate. The irony is that much more often it is actually the process of coordination that discovered other problems with the design in the first place. The word is used to abuse, and the contractor must be sensitive to such efforts to distort responsibility.

### 3.8.2 Operative Definition

1. Coordinating the work does *not* mean inventing information. It means:

   - Securing relevant information from those responsible for generating it
   - Fitting that information into the project requirements and confirming its appropriateness
   - Communicating that information to those responsible for finally approving its incorporation into the project
   - Distributing that information to all those requiring it, in order to allow them to complete their work at all those points where it interfaces with the information in appropriate time

   In other words, it means finding, assembling, and distributing information. It is *not* making up information, or fixing the mistakes in the specification when they're finally discovered. That's the designer's job.

2. Coordinating the work does *not* mean coordinating the design. Whenever there is an apparent conflict or confusion in the specifications, expect to be told flatly that it was *your* responsibility to "coordinate." The truth will more likely turn out to be that it was precisely your "coordination" that disclosed the problem in the first place.

**Example.** Two subcontractors both take exception to a certain item of work. After your own review, you conclude that each subcontractor actually reasonably inferred from their respective specification sections that the work would be done by the other subcontractor. The owner will therefore likely argue:

1. The work is specified *twice*; give me a credit, and
2. The GC should have "coordinated" the item; you determine who's going to do it, and just get it done.

The appropriate response will be based on ideas related to those in Section 3.5 *Specification "Interpretation,"* and on the principles outlined above. You will demonstrate that:

1. You *did* coordinate; that's what discovered the problem in the first place.
2. Each subtrade reasonably interpreted the specification to determine that the work was not included in their competitive bid (refer to Section 3.3.3 *"Intent" vs. "Indication"* and Section 3.4 *"Reasonable Review"*). The work was not, in fact, incorporated twice. It was not incorporated at all. A change order *is* therefore appropriate, but at an increase, not decrease, in cost.

Refer to Section 3.8.3 *Sample Letter to Owner Regarding Lack of Design Coordination* following as further example of proper treatment.

### 3.8.3 Sample Letter to Owner Regarding Lack of Design Coordination *(page 3.32)*

The Sample Letter to the Owner Regarding Lack of Design Coordination follows through on the ideas of Section 3.8.2 in the face of having discovered an error or conflict in the contract documents and having been accused of failing to "coordinate." Specifically, it:

1. Describes the conflict in as concise terms as possible.
2. Indicates directly how each side of the conflict was interpreted, and that it was reasonable to so interpret.
3. Clarifies that it is the GC's responsibility to coordinate the *work*—not to coordinate the design.
4. Confirms that you *did* coordinate; that's what disclosed the designer's failure to coordinate the design.
5. Points out that the design flaws are not your problem and that you're entitled to a change order to cover the extra cost.

Additionally, you might consider explaining that had there been no such confusion in the documents, the item of work would have been properly interpreted and its cost incorporated into the original bid. Its additional cost now is therefore no real embellishment of the contract price.

## 3.9 The Schedule of Values

### 3.9.1 General

This section discusses principles of the Schedule of Values and appropriate application in terms of identifying price structures to be used throughout the payment processing procedures and controlling would-be abuses during change pricing. The actual procedures for assembling, correlating, and communicating all this information for both the subcontractors and in the general process to the owner is described in detail in Section 4.

Refer also to Section 4.8 *Subcontractor Schedule of Values* for important related discussion.

### 3.9.2 Principles

The Schedule of Values is for partial payment invoicing processing *only:*

1. It is *not* a schedule of contract unit prices.
2. It is *not* to be used as any basis for adjustment in the contract sum.

**Example:** A schedule of values typically will include breakdowns of quantities and unit prices. If it is eventually determined that actual quantities for an item finally total to a number less than that shown on the schedule, the owner cannot adjust the line-item price downward. That kind of treatment is only appropriate where items are clearly part of specific unit price, allowance, or alternate contract price qualifications.

## 3.8.3
## Sample Letter #1 to the Owner Regarding Lack of Design Coordination

**Letterhead**

(Date)

To: (Owner)

CERTIFIED MAIL
RETURN RECEIPT REQUESTED

RE: (Project)
(Company Project #)

SUBJ: Design Coordination

Mr. (Ms.) ( ):

We agree that it is our responsibility to coordinate the work. It is not, however, our responsibility to coordinate the documents.

Per my letter of (date), the work (brief description) is actually described in both Sections ( ) and ( ) to be provided in the "other" section. Accordingly, neither subcontractor carried the cost of the item.

After reviewing the specification sections you reference, it is apparent that both subcontractors were justified in their views. Having made such reasonable interpretations, it is neither the subcontractors' nor the general contractor's responsibility to complete a search of documents to discover such mistakes in them; that is the designer's responsibility.

Moreover, it is the owner's responsibility to properly interpret the documents per General Conditions Article ( ). If you feel that the work has, in fact, been properly specified, it is your responsibility to provide the precise specification section. If you cannot, or for any reason will not, we will have no basis to direct the subcontractor to perform its "plans and specifications" work.

Accordingly, please confirm that a change order will be processed to cover the cost of this work.

Very truly yours,

COMPANY

Project Manager

cc: Architect
File: Change File ( )
CF

### 3.9.3 Level of Detail

The schedule should be broken down into at least a sufficient amount of detail that will allow clear demonstration of the actual completed work. Many individuals resist providing great amounts of detail in the schedule of values. The fact, however, is simply that the greater the level of detail in the schedule, the easier it will be to consistently substantiate your billings.

Prior approval of the detailed schedule of values removes to that degree an amount of subjectivity inversely correlated with that level of detail. If, for example, a detailed schedule is approved at the beginning of the project, simply having the percent complete of the individual line items approved in the regular payment procedure gets their values approved automatically.

In contrast, an oversimplified schedule will have fewer, but larger values for each general item. The larger the respective value, the more difficult it will be to justify in terms of percent complete in a given period. The final irony is that in order to actually justify billings on a large lump-sum item, you'll have to make up a detailed breakdown anyway. This will be an unofficial, unapproved breakdown that will then be subjected to intense scrutiny and debate.

Avoid the problem. Have your schedule of values in proper level of detail approved at the project's onset, and thereby keep your billing reviews smooth and without major incident.

### 3.9.4 Defense Against Price Adjustment Attempts

When anyone tries to use the schedule of values either to inappropriately adjust the contract sum or to make it a basis to modify other prices (such as change orders), firmly reject the maneuver.

**Example:** It may be observed that a scheduled payment value is actually much higher than the item's actual cost, and the owner or architect objects to the payment. Argue first that construction risk is *not* line-item risk, but bottom-line risk. As such, it is likely that there are other items that will show the opposite effect. Accordingly, if the owner is insistent in the "win" approach to adjust *this* item, then that "win/lose" approach should then be applied to *all* items on the schedule.

Unless you're clearly under a unit price (and not a lump-sum) contract, it will be straightforward to point out that there is simply no contractual basis for any such adjustment.

## 3.10 Requisitions for Payment and Contract Retainage

### 3.10.1 General

This section discusses contractual considerations of the requisitions for payment and retainage. Refer to Section 4 for additional procedures for development, submission, approval, and management processes.

Refer also to Section 4.8 *Subcontractor Schedule of Values* for important related discussion.

### 3.10.2 Maintenance of Billing Accuracy

By contract, statute, or public policy, the general contractor will be obligated to release funds to subcontractors to the same manner and extent that those funds have been released to the general contractor by the owner on behalf of that subcontractor. To put it another way, if the owner approves 80% payment to the general contractor for concrete, the general contractor must then release 80% to the concrete subcontractor.

It should be clear to any prime contractor, then, that if the concrete subcontractor, for example, is not really 80% complete, it is *very* risky to allow yourself to overbill, intentionally or unintentionally, for the item; legally the extra money will only have to be turned over to the subcontractor, who will not be entitled to it. Add to this the idea that if the general contractor is bonded and the subcontractor is not, it will be the *general contractor, not* the subcontractor, who will be guaranteeing completion on the work to the owner. This is probably the strongest argument against allowing any subcontractor to be overpaid.

It is therefore crucial that the subcontractor's schedule of values be comprehensive, in sufficient detail, and equitable—with particular scrutiny given to the real value of the *last* 25% or so of the subcontract item. All eyes must be continually focused on the real value of *remaining work to be completed.* Otherwise if you intentionally or inadvertently get overpaid on a subcontractor's behalf, you will be setting yourself up to be responsible to overpay the subcontractor, and thereby increase the risk to the Company of exposure to incomplete work.

### 3.10.3 Correlation of Subcontractor Schedule of Values with the General Schedule of Values

It is too common for subcontractor schedules of values to the general contractor to be too loosely correlated with the general schedule of values to the owner. This may be in large part because the general schedule of values is one of the first project documents to be expedited to the owner. Therefore it is prepared and submitted before every subcontractor's schedule of values is received or approved, possibly even before major bid packages are finalized.

If this is allowed to happen, an inordinate amount of subjectivity will be introduced into the process that will make correlating the amount due subcontractors with those amounts actually received on their behalf from the owner quite a challenge.

Avoid the arguments. Establish a procedure in the general schedule of values submission to the owner in which the values of major bid packages that have not yet been resolved are left as single lump-sums in the first submission. When the respective schedules of values are received and finalized for the subcontractors, the lump-sums on the general requisitions will be supplemented with this detail that will remain in that form for the duration of the project. Done in this way, early billings for the first items of work in process will not be delayed because of a late submission of the general schedule of values that was waiting for subcontractor breakdowns, and all eventual billings for each subcontractor will correlate on a one-to-one line-item basis with the general schedule of payments to the GC.

### 3.10.4 GC/Prime Contractor Retainage

The amount to be withheld from the general contractor as retainage from each payment is normally specified in the general conditions. Understand, however, that federal and state statutes dictate public policy regarding the actual amounts allowed to be withheld on public projects. If your contract, for example, states that

10% will be withheld, statute, or even the regulations of the funding source may require, for example, that only 5% be withheld. The regulations determining the procedure may be different even among agencies of the same state. Projects for the State of Connecticut Department of Housing, for example, were requiring 10% retainage during the same time period where the state's Department of Public Works required $2\frac{1}{2}$%.

If the retainage amount on your contract seems excessive, check with the relevant laws for a possible source of relief.

### 3.10.5 Limits on Subcontractor Retainage

There is likely to be some statutory provision in each state that is intended to regulate the maximum amount of retainage that can be held by a general contractor on a subcontractor. Many states, for example, may provide that the general contractor can only withhold the same percentage as is being held on the GC by the owner. Even where it cannot be shown that there is a clear statement in existence, astute subcontractors may attempt to apply the Pass-Through Clause (see Section 3.3.8) to make the general percentage value apply. They will accordingly argue, for example, that if the owner is holding 5% on the GC, the GC may not withhold 10% on the subcontractor.

In these cases, it is important to realize that there may be circumstances that materially alter the justification for such an interpretation.

**Example:** If the general contractor has 100% payment and performance bonds to the owner, and the subcontractor has no such bonds to the general contractor, the clean pass-through relationship has been materially altered. The additional retainage may be considered an appropriate amount of needed security for faithful performance to help make up for the lack of the subcontractor's bonds.

### 3.10.6 Substitution of Securities for Retainage

Although it is likely that this option will not be mentioned anywhere in the contract documents themselves, many states provide for the ability of the contractor to substitute certain marketable securities in exchange for release of all or part of retainage to the value of those securities.

If, for example, you have money tied up in an interest-bearing vehicle such as a Treasury Bill, that Bill might be given to the owner (state) to hold as its security for your faithful performance in lieu of retainage. Plain retainage normally does not bear interest. In so allowing the corresponding amount of retainage to be released, you effectively improve your cash flow to the level of the retainage released. Another way of looking at it is that you've substituted interest-bearing retainage for non-interest-bearing retainage.

Check the laws of your state and the list of acceptable types of securities.

## 3.11 Acceleration of Work

### 3.11.1 General

If the Owner should direct the contractor to complete the work earlier than the originally required date, it will be fairly obvious that the contractor has been accelerated. In such a clear situation, it is easy to see how the contractor will be entitled to recover the costs of such acceleration under the change clause for the cost of such acceleration, plus overhead and profit.

A less obvious situation occurs when the contractor is required by the owner to complete the work by the original contract date when there has been excusable delays that entitle the contractor to an extension of the original contract time. Depending upon the category of the otherwise excusable delay, the contractor may also have been entitled to recovery of costs associated with the time extension. This type of action is a *constructive acceleration,* and should entitle the contractor to a change order to cover the increased cost.

### 3.11.2 Justifications for Constructive Acceleration

There are minimum requirements generally used by courts as a test of whether the contractor has been constructively accelerated:

1. There must be an excusable delay for which the contractor is entitled to an extension of time.
2. The contractor must have requested an extension of time (with or without associated costs).
3. The owner must have refused or failed to grant the extension of time.
4. The contractor must have been required to complete the contract without an extension. Note that this can be either expressly or impliedly required either directly or by the conduct of the owner's authorized representatives.
5. The contractor must have actually completed the contract on time and actually incurred the extra costs.

If all these conditions are present, the contractor should be entitled to recover its costs, plus reasonable overhead and profit: an equitable adjustment in the contract sum.

Remember also the Pass-Through Clause (Section 3.3.8): these same principles of constructive acceleration apply in the same way between the general contractor and the subcontractors.

**Example:** The government temporarily suspends work on a public project for reasons not the fault of anyone on the construction force, but it then requires completion by the original contract date. If all the notification provisions as outlined above have been met, affected subcontractors will experience excusable delays which will be chargeable to the general contractor. The general contractor must accordingly assemble all subcontractor constructive accelerations into the complete package that will represent the total constructive acceleration costs of the general contractor and construction force to the owner.

### 3.11.3 Types of Recoverable Acceleration Costs

Different types of costs, some of which have been much more obvious than others, have been held to be recoverable in different cases. Examples of the more obvious costs include:

1. Overtime wages
2. Additional supervision
3. Expediting efforts (such as travel to a supplier's fabrication location to shorten shop drawing or coordination time)

4. Costs associated with changes of construction sequence that may alter material storage/handling, temporary protection, or other direct costs relating to coordination (the elevator's not working yet, so everything had to be carried...)
5. Higher material prices caused by special fabrication cycles
6. Higher shipping/transportation costs

It's usually a straightforward procedure to determine such direct and more obvious costs. Simply visualize the work and identify all those activities to which you are now subjected that would not have been required if the acceleration had not occurred. This is the basic concept of "equitable adjustment." An example of a less obvious cost associated with direct or constructive acceleration is the loss of normal labor efficiency caused by changes in scheduling and the natural loss associated with working longer hours.

It is an accepted fact in manufacturing circles, for example, that with all other things being equal with the exception of time-of-day, the second shift for varying reasons will be a certain percent less efficient than the first shift. The third shift will be even less efficient than the second. The theories as to why are complex, but the fact is undeniable. The idea has been the subject of intense study over the years, which has resulted in a catalog of generally accepted efficiency rates for different manufacturing shifts.

The problem in construction is that when these same ideas are applied to a construction force in an acceleration condition, the factors are extremely difficult to quantify and prove. Every situation is different, with its own set of complex characteristics. Your success in demonstrating these efficiency losses will be directly related to the level of detail and the accuracy of your records.

Another example of a less obvious acceleration cost is the increased cost of performance caused by the impact or ripple effect on other work that had not been changed or otherwise directly accelerated. That work must now be altered or accelerated in order to accommodate the directly accelerated portion of the work.

Be aware of the problem associated with condition #4 of constructive acceleration in Section 3.11.2. The owner may stall or otherwise delay its response to the contractor's request for a time extension. If this happens, it may have the effect of forcing you to choose between voluntary acceleration or a breach of contract for late completion. In such a case, there will be too much subjectivity introduced into the process if you allow it, and you may be setting yourself up for an acceleration "directive" that cannot be proved—and therefore will not be recoverable. Accordingly, after a reasonable time has passed for the owner to respond to your request, push the answer by calling attention to the constructive acceleration. Consider the *Sample Letters to the Owner Regarding Constructive Acceleration* in Sections 3.11.4 and 3.11.5.

### 3.11.4 Sample Letter #1 to the Owner Regarding Constructive Acceleration *(page 3.39)*

The Sample Letter #1 to the Owner Regarding Constructive Acceleration follows through in one way on the suggestions of the preceding section. Specifically, it:

1. Calls attention to the previously identified and documented excusable delay

2. "Reminds" the owner that it has required the contractor to forgo the excusable delay and *accelerate* the work to bring the completion date back to the original time
3. Confirms that the time of requested owner confirmation has passed, leaving owner direction as to its wishes unclear
4. Requires *definite and specific* direction before any action is taken by the construction force to actually accelerate

### 3.11.5 Sample Letter #2 to the Owner Regarding Constructive Acceleration *(page 3.40)*

The Sample Letter #2 to the Owner Regarding Constructive Acceleration follows the same steps as Letter #1, with the exception that it makes the decision to definitely accelerate. It accordingly attempts to pin the owner to a concrete action that will later be viewed as a definite owner directive to constructively accelerate.

## 3.12 Liquidated Damages

### 3.12.1 General

The liquidated damages provision is often misunderstood in terms of its real potential impact on the project. Most times, it is perceived to be a heavy weight suspended from a string over the contractor's head that could snap at any moment. The reality is usually nowhere near this severity, and the truth on a particular project may wind up actually being a comfortable limitation on the owner's ability to make trouble.

It is crucial that you develop a complete perspective on the liquidated damages realities on each individual project both at the project's start and whenever there is trouble potentially associated with a nonexcusable (or alleged nonexcusable) delay. Have a clear idea on both the owner's rights and abilities and those of each subcontractors' Pass-Through (see Section 3.3.8) responsibilities.

### 3.12.2 Definition

It is recognized that:

1. If the owner is delayed in its ability to occupy the project and use it for its intended purpose, it will suffer damage.
2. It is extremely difficult, time-consuming (and therefore very expensive), and very subjective to define the actual extent of those damages, and to distill those values to a daily assessment to be directly assigned to any delay.

The "liquidated damages" provision of the contract is therefore provided to simplify the whole thing: if such a delay occurs, damages on the part of the owner will be justified. It is nothing more than a simple stipulation of both contracting parties as to the value of daily damages to be assigned *if the owner is actually delayed through the fault of the contractor.* It is there *only* to avoid the argument of determining the *value* of damages.

## 3.11.4
## Sample Letter #1 to the Owner Regarding Constructive Acceleration

Letterhead

(Date)

To: (Owner)

CERTIFIED MAIL
RETURN RECEIPT REQUESTED

RE: (Project)
(Company Project #)

SUBJ: Constructive Acceleration

Mr. (Ms.) ( ):

On (date), we advised you of the specific conditions justifying an extension of time, and itemized associated costs.

On (date), Mr. ( ) notified us that you still expect the contract to be completed by its original date.

On (date), we accordingly requested that you confirm either your acceptance of the time extension as presented, or provide us with your written directive to accelerate the work in order to meet the project's original completion date. To date, we have no return correspondence.

Please be advised that until we receive such a proper directive to accelerate, we are continuing with the work in accordance with the schedule indicated in the (date) time extension request, in anticipation of your acceptance of the time extension as indicated therein.

Very truly yours,

COMPANY

Project Manager

cc: Architect
Anyone named in the letter
File: Contract Time
Sched. Rev. # ( )
CF

## 3.11.5
## Sample Letter #2 to the Owner Regarding Constructive Acceleration

**Letterhead**

(Date)

To: (Owner)

CERTIFIED MAIL
RETURN RECEIPT REQUESTED

RE: (Project)
(Company Project #)

SUBJ: Constructive Acceleration

Mr. (Ms.) ( ):

On (date), we advised you of the specific conditions justifying an extension of time, and itemized associated costs.

On (date), Mr. ( ) notified us that you still expect the contract to be completed by its original date.

On (date), we accordingly requested your direct confirmation of your order to accelerate, along with your agreement to pay for all costs associated with such acceleration. To date, we have no return correspondence.

Please be advised that your lack of response is now being construed as confirmation of your Mr. ( )'s directive to accelerate.

Accordingly, unless we receive in writing any change to this directive to accelerate, actions will be taken to accelerate the work in order to regain the original schedule. All costs associated with such acceleration, together with overhead and profit, will be consolidated and submitted as a change order.

Very truly yours,

COMPANY

Project Manager

cc: Architect
Anyone named in the letter
File: Contract Time
Sched. Rev. # ( )
CF

**Example #1:** A road project is delayed a month. Is the public "damaged" because of its inability to use the road? If so, how much is the damage worth? Is the Public Agency administering the contract damaged? Why? After all, isn't its funding tied to local politics?

**Example #2:** A commercial client is delayed in occupying its space. Lost rent may be easy to calculate, but in this market would they really have rented the space at all? If it is a manufacturing, sales, or distribution organization, what is the value of not being able to ship orders for that time period?

Liquidated Damages does not *justify* the concept of damages, does not make them in any way automatic, and does not in any way make it any easier for the owner to prove or actually collect damages. It is an arithmetic convention designed to avoid only the argument of defining the actual value of any damages.

### 3.12.3 Concepts and Clarifications

Several concepts and clarifications are necessary in order to properly apply the principles of this section:

1. *"Liquidated Damages" is not a "Penalty."* The term "Liquidated Damages" is used first specifically to avoid the use of the word "penalty." It has, for example, long since been interpreted by the courts that if a contractor is working under threat of a "penalty" for failing to complete the project by a prescribed date, then the contractor would be correspondingly entitled to a "reward" to the same extent if the contract is completed *prior* to that date. What's good for the goose is good for the gander. The use of the word "penalty" is therefore avoided in order to preclude the contractor's ability to collect any such "reward."
2. *"Liquidated Damages" is not an automatic cost that occurs after the original contract end date has passed.* Liquidated Damages cannot be assessed against the contractor for delays incurred which fall within the scope of any excusable delay clause—for delays demonstrated to be not the fault of the construction force. The owner is just as responsible to prove its case of justification for any damages in the first place.
3. *The presence of a Liquidated Damages provision may in some cases actually improve your position to negotiate or settle problems.* Because the amount set forth in the contract as the value of damages has the effect of *limiting* the owner's ability to assign more than that amount, it immediately controls the size of threats and restricts posturing and legal maneuvering designed to puff up the size of claims or counterclaims by the owner. You know exactly the maximum size of any potential owner claim—the owner can't so easily quantify yours.

### 3.12.4 Technical Defenses and Considerations

1. *Apportionment.* If *both* the contractor and the owner have contributed to the cause of the total delay, the contract may provide that the responsibility be "apportioned" between the contractor and the owner. Liquidated Damages will in this way be assessed against the contractor for its portion of the delay, and the contractor would be entitled to prove its damages for the owner's portion of the delay.

   If the contract does not provide for such apportionment, some courts have held that, by contributing to the delay, the owner totally waives any claim to Liquidated Damages. Other courts, however, have held that the contractor is only entitled to a "credit" for that part of the delay caused by the owner—another way of saying "apportionment."

2. *Application in Contract Abandonment.* If a contractor breaches a contract by totally abandoning performance, some courts have determined that the Liquidated Damages provisions may not apply. These courts distinguish between a breach by delay and a breach by abandonment, finding that the parties intended to apply the Liquidated Damages only to delays when the contractor completes the work.
3. *Application after Substantial Completion.* Courts generally refuse to allow Liquidated Damages if the contractor's performance under the contract has been *substantially completed* by the date specified in the contract and only punchlist or detail work remains. Substantial Completion has generally been defined in a manner consistent with that used by the American Institute of Architects as:

   ...the date certified by the architect when construction is sufficiently complete in accordance with the contract documents, so the Owner can occupy or utilize the Work or designated portion thereof for the use for which it is intended.

   It is therefore generally recognized that when the owner has possession or use of the property, an assessment of Liquidated Damages for trivial details in the work would amount to a penalty.
4. *Application in Cases of Partial Occupancy.* If you've built or renovated a school, for example, and 80% of the facility had been accepted and occupied by the date stipulated in the contract, is the owner justified in assessing 100% of the value of Liquidated Damages?

   In these types of cases, it has been successfully argued that by its definition, the concept of Liquidated Damages can only apply when the entire project is delayed. In the case of partial occupancy, therefore, is it inappropriate to assign even a portion of Liquidated Damages?

   Be extremely careful on this one. If the food service equipment, for example, is the item of incomplete work, and if you successfully argue that Liquidated Damages does not apply, you may be opening yourself up for an assessment of *actual* damages. The school, for example, now has to arrange catering, and it has a food service staff doing little or nothing for which they still have to be paid. You may be better off negotiating a percentage of Liquidated Damages if it comes to that.

## 3.13 Guaranties and Warranties

### 3.13.1 Definition of Terms

The words "Guaranty" and "Warranty" have different meanings, and they should not be construed as being interchangeable.

Although the two terms are often characterized as being the same, in strict legal usage, the terms are different. A *warranty* has been defined as an absolute liability on the part of the warrantor, and the contract is void unless it is strictly and literally performed. A *guaranty* is a promise not imposing any primary liability on the guarantor, but binding him or her to be answerable to the default of another.

In other words:

A *Warranty:*

- Binds a party to the terms of its contract
- Implies present or future liability
- Has no time limitation on the liability if "unrestricted"
- Applies usually to products and their qualities

A *Guaranty:*

- Binds a third party to the terms of another's contract (*Example:* Performance Bond)
- Implies present or past liability
- Has a stated time limit usually
- Applies to indebtedness or performance of one's duties

### 3.13.2 Date of Beginning Coverage

General contracts usually require guaranties and warranties to begin from some stage of contract completion at which the project can be used for its intended purpose, i.e., "substantial completion." Be sure that all warranties list the term "substantial completion" as well as the corresponding actual date as the start of the warranty period.

Refer to the respective specification section, general conditions of the particular contract, and other subcontract provisions to finalize complete requirements and other necessary, specific language.

Most subcontractor and supplier guaranties and warranties will be required to be countersigned by the general contractor prior to delivery to the owner. Verify all such requirements.

### 3.13.3 Express vs. Implied Warranties

1. *Express.* Many products carry a standard written (express) warranty that:

   *a.* Is usually in force for a prescribed period (often one year) from date of *shipment*

   *b.* Does not cover damages in shipment or abuse by other trades before acceptance

   It can be difficult to get a distant product manufacturer to provide for all the specific components of a required warranty after the order has been placed. Even if the sub-trade responsible will positively make up the difference in such warranties, these supplements may not be accepted by the owner.

   Be sure that the complete product warranty provisions are part of the purchase agreement in the first place—either directly to that vendor or passed through the affected subcontractor.

2. *Implied.* There is usually an implied warranty in every transaction or contract, unless the documents clearly state otherwise, that requires the product or goods to be reasonably fit and sufficient for the purpose for which they are to be used.

*Example:* When a product is supplied in response to a performance specification (see Section 3.3.6), there is an implied warranty that the product will perform as required for that purpose.

*Example:* When an owner requires a contractor to build in accordance with plans and specifications provided or prepared by an architect or engineer, the owner impliedly warrants that those plans and specifications will be sufficient to achieve the desired results, that the project *can* be built as designed, and that all specified materials *can* be ordered and delivered within the time necessary.

If you cannot get your specified materials in time because they were not available through no fault of yours, the *designer* will have violated *his/her* implied warranty of availability of such products within the time required to complete the contract.

## 3.13.4 Submission Dates

1. The subcontract or purchase order is the first document that requires a respective sub-vendor to provide guaranties and warranties in accordance with the requirements of the contract documents. The *Sample Form Letter to Subcontractors Regarding Submittal Requirements* (Section 4.10.3) is the next early warning reminder.
2. The *acceptable* date of substantial completion must be confirmed with the owner. If this is not done in advance, you will run the serious risk of spending all that time and energy securing guaranties and warranties with unacceptable dates. They'll all have to be redone.
3. When the date of substantial completion has been so confirmed, all responsible vendors must be notified that:

   *a.* Immediate submission of the document(s) is required, *in proper form.*
   *b.* Final payment cannot be released without such proper, timely submission.
   *c.* Any damages resulting from delays in final payment to the general contractor directly related to the failure to provide such proper, timely submissions will be charged accordingly.

Refer to the *Sample Sub-Trade Guaranty/Warranty Notification Letter* of Section 3.13.6 to follow through on the procedure.

## 3.13.5 Form

1. If the contract documents are complete, they will have either the specific language and form for each respective guaranty and warranty delineated or a direct reference to the correct document. In such cases, it is a straightforward matter to define the requirements and enforce compliance. Beyond that, it must also be confirmed that any requirements beyond those in the project documents that have been added by way of subcontract or purchase order must also be confirmed.
2. In the absence of a specifically required form, request one from the owner that will be acceptable. Review such a form for adherence to the contract without embellishment.
3. It is common, but not automatic that sub-trade guaranties and warranties are to be countersigned by the general contractor. Do not countersign if it is not specifically required.

### 3.13.6 Sample Sub-Trade Guaranty/Warranty Notification Letter *(page 3.46)*

The Sample Sub-Trade Guaranty/Warranty Notification Letter is designed to compel all responsible parties to submit what is usually one of the last requirements of their contracts. It is a form letter that follows through on the suggestions of this Section 3.13 by:

1. Requiring timely submission, in proper form
2. Advising of the date that the guaranty or warranty is to take effect
3. Advising that payments will be affected until the proper submission is received
4. Putting the party on notice that problems directly or indirectly resulting from delay in the proper submission will be charged

Have a supply of letters on hand to be filled out and sent by the project manager or project engineer immediately after receipt of an official correspondence from the owner confirming the date of substantial completion.

Refer also to Section 4.18 *Securing Subcontractor/Supplier Guaranties/Warranties* for important related discussion.

## 3.13.6
## Sample Sub-Trade Guaranty / Warranty Notification Letter

| Letterhead |
| --- |

Date: ______________________

To: ____________________________
____________________________
____________________________

Confirmation of Fax:
Fax #: ____________________

ATTN: ____________________________

RE: Project: ____________________________
Company Project #: ____________________

SUBJ: Guaranties / Warranties: Section(s) __________

Mr. (Ms.) __________________________:

The Substantial Completion Date of the Project is ______________________, 19 ____. As you are aware, this is the date on which your guaranties and/or warranties as required in the subject specification section(s) take effect.

Submission of your guaranties and/or warranties *in their specific, proper, and complete form* is required by ______________________, 19 ____ in order to allow the current payment to be processed.

Failure to provide your correct guaranties and/or warranties by the above date will delay the submission of the general requisition to the owner that includes retainage reduction and final payment to the entire construction force.

Please don't allow yourself to be in the position of being responsible for delay in such payment: submit your proper documents as requested. Your prompt compliance is crucial.

Please call if you require further clarification.

Very truly yours,

COMPANY

Project Manager

cc: Accounts Payable
File: Final Payment
Sub-Vendor File: ________
CF

Section

# 4

# Project Engineering

## 4.1 Section Description

Project engineering is where the administration of construction is done. In a real sense, it is the how-to of building the project on paper—before it is built in the field.

The fit of the project engineering function within the organization is outlined in Section 1.5. Here we provide the step-by-step activities that are necessary to actually fulfill the Project Engineer's complex responsibilities. Sample full-size forms, checklists, flow charts, word-for-word letters, and specific instructions are provided at each step of the way.

This section is very closely coordinated with:

- Section 5, *Site Superintendence*
- Section 6, *Safety and Loss Control*
- Section 7, *Progress Schedules and Funds Analysis*

Although this section includes those specific activities that fall within the restricted definition of the project engineering function, actual practice will necessarily dictate large amounts of overlap of individual responsibilities across these disciplines.

## 4.2 Responsibilities of the Project Engineer

The specific assignments of the actual project staff will vary greatly from organization to organization and from project to project. The actual activities detailed in this section may, in practice, be divided among several people and even to a large extent redundant to the activities of other individuals—spread across the responsibilities of the Project Manager, Project Engineer, Site Superintendents, Scheduler, and any other staff enlisted to support these functions.

Refer to Section 1.5 for a summary overview of the project engineering responsibilities and to the table of contents of this section for specific related activity lists.

## 4.3 Project Files

### 4.3.1 General

The establishment of the project files is treated in detail in Section 2.4, *Files and File Management.* It is included in that section in order to allow those with responsibility for other portions of this Manual to be aware of the complete file structure system, and to coordinate their activities with the rest of the organization.

### 4.3.2 File Responsibilities of the Project Engineer

In referring to Section 2.4, the Project Engineer will be the one responsible to prepare and maintain:

- The plans, specifications, and changes (2.4.3)
- The General Project File (2.4.4)
- The Clarification/Change Log (2.4.5)

- The Subcontractor Summary and Phone Log (2.4.6)

In this way, each of the documents generated in this section will be properly consolidated with the complete project record.

### 4.3.3 Duplicate Files—Home and Field Offices

Certain files will be duplicated between the home office and the field, modified as indicated hereafter. Those include:

1. *Plans, Specifications, and Changes* (as described in Section 2.4.3). Both the field and office sets must be maintained and posted with all changes, modifications, clarifications, etc., on a current basis, and must be precise duplicates of each other.
2. *The General Project File.* The General Project File at the home office will be complete in every respect, as detailed in Section 2.4.4. The duplicate file in the field will be set up and maintained with the following modifications:

   *a.* Documents relating to solicitation, bid package evaluations or reviews, and subcontract negotiations will not be distributed to the field.

   *b.* Field copies of final, executed purchase orders and subcontracts will have prices omitted for security reasons. This information will be available to the field staff as necessary, but no hard copy is to be left on-site.

   *c.* Assuming that shop drawing review by the project engineering function is being conducted at the home office (off-site), copies of submittals for approval will not be sent to the field during the approval cycle unless unique circumstances require an expediting effort conducted from or with the field. This would be a special circumstance to be managed with the utmost caution.

   Instead, the field will receive copies of all submittal transmittals so that all are aware of the current status of particular approvals. Only copies of "approved," "approved as noted," or equivalent (see Section 3.6, *Shop Drawing "Approval"*) that are actually to be used for construction are to be distributed to the field file.

   Beyond these qualifications, all other project correspondence and documents are to be distributed to the field on a current basis, noted with the same file instruction as for the home office.

## 4.4 Subcontractor Summary and Telephone Log

### 4.4.1 General

Immediately after the award of a contract, prepare a 1-inch three-ring binder labeled *Subcontractor Summary.* On small to mid-sized projects, or projects that otherwise will not have relatively large numbers of approval submittals, this binder could be combined with the *Submittal Log* of Section 4.9.

This binder will be the location in which the summaries of each Subcontractor will be noted and performance compliances (and noncompliances) consolidated daily. It will provide at a glance the complete status on a current basis of each Vendor's actions and, more importantly, their specific deficiencies.

### 4.4.2 Purpose

A project will have dozens of Vendors. Each Vendor will have dozens of individual requirements, including, for example:

1. Submissions required by the project documents, such as:
   - Vendor approval by the Owner
   - Complete submission of shop drawings and samples
   - Periodic submission of certified payroll reports
   - Delivery of lien waivers for itself and for its Vendors
2. Submissions required by its Subcontractors such as:
   - Delivery of insurances in proper form
   - Certified list of Sub-Sub-Vendors
3. Adherence to specific subcontract requirements, such as:
   - Confirmation of Schedule requirements
4. Responses to or confirmation of receipt of Company forms and special correspondence sent periodically, such as:
   - Submittal Requirements Letter of Section 4.10
   - Backcharge Notifications of Section 4.16
   - Construction Schedule transmittals and notices

These add up to hundreds of time-sensitive issues and requirements that must be policed by the Project Engineer on a current basis for each project to ensure that each Vendor is continually being brought into compliance.

The purpose of the *Subcontractor Summary* and the *Telephone Log* is therefore to:

1. Provide a vehicle that will streamline the Project Engineer's review of each Subcontractor's compliance in terms of both performance and all the various submissions and compliances of their subcontracts
2. Provide the space for a phone log that will consolidate very powerful information both for daily subcontract management situations and for backup in a dispute

Together, the two components provide a quick but complete history on the total performance of the party. This will permit exerting constant pressure on the process to close all those cracks through which things can fall too easily.

You'll be able to substantiate to the minute:

- Chronically necessary expediting efforts on your part
- Chronically missed dates or other performance failures
- Specific reasons why payments should be withheld, with all the backup
- Direct responsibilities for delays and interferences
- Substantiation of all backcharges, along with all those chances given the other party to avoid them

### 4.4.3 Preparation

1. Arrange alphabetical tabs in the 1-inch binder.
2. A *Subcontractor Summary Form* (see Section 4.4.6) is inserted alphabetically for each Subcontractor and Supplier.
3. Place three or four pages of the *Vendor Telephone Log Form* (see Section 4.4.8) immediately behind each Subcontractor Summary Form. Add to the supply during the project if it becomes necessary.
4. As other Vendor summary documents become available (such as the *Subcontractor / Supplier Reference Form* of Section 4.5), place the completed form directly behind the Subcontractor Summary Form.
5. The front of the binder is reserved for telephone logs for the Owner, the Architect, and other design professionals.

### 4.4.4 Procedure and Use

1. Use the standard form. Add any specific requirements unique to the project (such as special approvals), and photocopy a supply of those *project* forms.
2. Add any requirements specific to a particular Subcontractor or Supplier (a bond or special insurances, for example) to that Vendor's form. Do this *as* each subcontract or purchase order is being prepared.
3. *The most important idea is to develop immediately the habit of always having the binder open in front of you.* This aspect is crucial to using the power of this simple procedure to its fullest potential.
4. Use the binder as your regular telephone and address directory. In this way you'll automatically have it open to the Vendor as the call is being made, or as the form letter is being prepared.
5. When you are phoned by any Vendor, have the binder open to that Vendor's summary *before you pick up the phone.*
6. Log all information *as it occurs.* Since the binder is already open, it takes less than a moment, for example, to note the date of the summary as the Subcontractor/Supplier Delivery Requirements Form Letter of Section 4.11.3 is prepared.
7. Enter *all* information, especially when nothing has happened. If, for example, all you get for your effort is "he's not in; he'll call you back," note those remarks along with who gave them, date, and time. A list of six of these in a row, for example, goes a long way in demonstrating the performance failures of the other party.
8. Using the telephone log in this way will consolidate all information by date *per Vendor,* making the information immediately clear, available, and usable. This is in sharp contrast to including these kinds of notes, for example, on the Daily Field Reports. Records kept in that way will require hunting through the entire set every time a comprehensive summary of an effort is needed for a Vendor. It's just too much work, which will never get done, or done properly.
9. Subtly make everyone aware of these records. Having the records is the first step to getting the upper hand in every situation every time. If the other parties are *aware* of the existence of such precise, detailed records, they'll learn fast that it is in their best interest to avoid games. When they see your ability to greatly simplify the usually extreme complexity of project relationships, they'll take another look, and decide that it's best not to fool with *this* relationship.

Some ways to accomplish this include:

*a.* Whenever a Subcontractor phones to question, to complain, or for any other reason:

(1) Review that Subcontractor's performance summary before you pick up the phone.
(2) Before the party gets a chance to say what his or her problem is, fire off *every* piece of paper that he or she has failed to deliver properly. After that, advise of all those dates (and times) that you've been trying to reach him or her about *your* problem. Then go through *your* issues. When this process has been completed, *then* you can ask the caller what *his* or *her* problem is.

*b.* In any formal or informal meeting:

(1) Open the binder whenever a relevant issue is raised.
(2) Let everyone *see* you're making notes in their file *as the discussion is occurring*. This action gives profound legitimacy at a later date to *your* information as being the *right* information. It will make everyone think before they speak, and force them to consider their actions (or inactions) very seriously.

10. If more elaborate conversation or an impromptu meeting has taken place, consider using, in addition to your notes, the *Meeting/Conversation Record Form* of Section 4.13.12.

### 4.4.5 Sample Subcontractor Summary Form—Completed Example *(page 4.9)*

The Sample Subcontractor Summary Form that follows is the form to be used to follow through on the procedure described in this section. A completed form is provided as an example of its proper use. Section 4.4.6 provides the blank form to be photocopied for actual use.

### 4.4.6 Sample Subcontractor Summary Form—Blank Form *(page 4.10)*

### 4.4.7 Sample Telephone Log Form—Completed Example *(page 4.11)*

The Sample Telephone Log Form that follows is the form to be used to follow through on the procedure described in this section. A completed form is provided as an example of its proper use. Section 4.4.8 provides the blank form to be photocopied for actual use.

### 4.4.8 Sample Telephone Log Form—Blank Form *(page 4.12)*

## 4.5 Subcontractor/Supplier Reference Form

### 4.5.1 General

The Subcontractor/Supplier Reference Form is to be sent to each Subcontractor and Supplier along with their respective subcontracts or purchase orders. The

# 4.4.5
# Sample Subcontractor Summary (Completed Example)

PROJECT TRADESMAN SCHOOL
OWNER No. BI-Q-277 (B)
COMPANY No. 9900
SUB No. —

SUB FOUNDATIONS CORP.
292 AMITE BLVD
COLCHESTER, CT 06920
NAME JERRY GRANT

SECTION(S) 03200
03300
PHONE 203-624-1820
FAX 203-624-1841

1. Prefile Sub — X No ____ Yes; #____
2. Owner Selected Sub — X No ____ Yes
3. Subcontractor Approval Required — ____ No X Yes; Date Submitted 1-7-94; Date Approved 2-4-94
4. Subcontract — Sent 2-7-94; Received 2-19-94
5. Plans & Specifications — Sent 1-8-94
6. Information Request Form — Sent 2-7-94; Received 2-19-94
7. Schedule of Values Letter — Sent 2-7-94; SOV Rec'd 2-19-94
8. Submittal Requirements Letter — Sent 2-7-94
9. Preliminary Construction Schedule — Sent 2-7-94; Comments Rec'd ____
10. Final Construction Schedule — Sent ____
11. "Copy" Notice — Sent ____
12. Required Material Deliveries: Item: NONE; Date: ____
13. Scheduled Installations/Erections

| Item | Duration |
|---|---|
| FOOTINGS | 25 W.D |
| FOUNDATIONS | 30 W.D. |
| PIERS | 10 W.D. |

Correspondent: JERRY GRANT
Remarks/Qualifications: FOUNDATIONS START (10) W.D. AFTER FOOTING START. PIERS PLACED W/IN SAME TIME FRAME AS FOUNDATIONS

14. Materials Received: ____ Ahead of Schedule: Date: ____
____ On Schedule
____ Behind Schedule: Date: ____
Remarks: SUBCONTRACT FOR LABOR-ONLY

15. Guaranties/Warranties — Requested ____ Received ____
16. As-Built Documents — Requested ____ Received ____
17. O&M Manuals — Requested ____ Recieved ____
18. All Closeout Documents — Requested ____ Received ____
19. Lien Waivers — Requested ____ Received ____
20. Final Release Form — Requested ____ Received ____
21. Backcharge Summary

| Date | Description | Amount |
|---|---|---|
| | | |
| | | |
| | | |
| | | |

22. Conversations, Notes, & Remarks (Continue on additional sheets as necessary)

| Date | Time | Description |
|---|---|---|
| 2-11-94 | 2:00 PM | JERRY ADVISED AMC IN TELECON THAT EXECUTED SUBCONTRACT WILL BE RETURNED IMMEDIATELY & 3-15-94 START LOOKS GOOD |
| 3-1-94 | 9:00 AM | TELECON; JERRY ADVISED PAUL D. THAT 3-15 MAY MOVE TO 3-16. |

## 4.4.6
## Sample Subcontractor Summary

PROJECT________________ SUB________________ SECTION(S)__________
OWNER No.________________ ________________ __________
COMPANY No.________________ ________________ PHONE __________
SUB No.________________ NAME________________ FAX __________

1. Prefile Sub ______ No ______ Yes; #______
2. Owner Selected Sub ______ No ______Yes
3. Subcontractor Approval Required ______ No ______ Yes; Date Submitted__________
   Date Approved __________
4. Subcontract Sent __________ Received __________
5. Plans & Specifications Sent __________
6. Information Request Form Sent __________ Received __________
7. Schedule of Values Letter Sent __________ SOV Rec'd __________
8. Submittal Requirements Letter Sent __________
9. Preliminary Construction Schedule Sent __________ Comments Rec'd __________
10. Final Construction Schedule Sent __________
11. "Copy" Notice Sent __________
12. Required Material Deliveries: Item:__________ Date:__________
    __________ __________
    __________ __________
13. Scheduled Installations/Erections Item __________ Duration:__________
    __________ __________
    __________ __________
    __________ __________

Correspondent:________________
Remarks/Qualifications:________________________________
________________________________________________
________________________________________________

14. Materials Received: ______ Ahead of Schedule: Date:__________
    ______ On Schedule
    ______ Behind Schedule : Date:__________

Remarks:________________________________________
________________________________________________
________________________________________________

15. Guaranties/Warranties Requested__________ Received__________
16. As-Built Documents Requested__________ Received__________
17. O&M Manuals Requested__________ Recieved__________
18. All Closeout Documents Requested__________ Received__________
19. Lien Waivers Requested__________ Received__________
20. Final Release Form Requested__________ Received__________
21. Backcharge Summary

| Date | Description | Amount |
|---|---|---|
| | | |
| | | |
| | | |
| | | |

22. Conversations, Notes, & Remarks (Continue on additional sheets as necessary)

| Date | Time | Description |
|---|---|---|
| | | |
| | | |
| | | |
| | | |
| | | |

## 4.4.7
## Sample Telephone Log Form (Completed Example)

**RECORD OF TELEPHONE CONVERSATIONS** Project: SOUTHINGTON STA. # 9890

| Date | Time | Discussion |
|---|---|---|
| 4.14.94 | 9:00 AM | M. JEFF CALLED PAUL GABRIEL (WESCON) TO REVIEW |
| | | ITEMS 7 & 12 ON THEIR SUBMITTED SCHED. OF VALUES. |
| | | PAUL OUT 'TILL NOON; WILL RETURN CALL. |
| 4.15.94 | 8:30 AM | M. JEFF CALLED PAUL BACK RE: 4.14.94 CALL; NOT IN |
| | | YET- WILL CALL BACK |
| | 1:10 PM | M. JEFF CALLED PAUL AGAIN; STILL NOT IN |
| | 2:40 PM | PAUL GABRIEL CALLED BACK; M. JEFF REVIEWED THE |
| | | S.O.V.; PAUL WILL RESUBMIT TODAY. |
| 4.18.94 | 11:05 AM | PAUL ADVISED AMC IN TELECON THAT ALL SHOP DWGS. |
| | | WILL BE DELIVERED TO JOBSITE ON 4.22.94 |
| 4.21.94 | 10:30 AM | M. JEFF CALLED PAUL GABRIEL TO CONFIRM SHOP DWG. |
| | | DELIVERY. PAUL SAID "THEY'RE COMING BACK FROM |
| | | THE PRINTER ON 4.22; WILL BE DELIVERED 4.23.94 |

## 4.4.8
## Sample Telephone Log Form

RECORD OF TELEPHONE CONVERSATIONS Project: ______________________#________

| Date | Time | Discussion |
|---|---|---|
| | | |

Vendor will be required at that time to complete the form and return it immediately to project engineering's attention.

### 4.5.2 Purpose

The purpose of the form is to greatly facilitate present and future communications. The form will:

1. Inform the Vendor of the Company project identification and job number that is to be used on all correspondence.
2. Solicit and confirm the names and phone numbers of all contact persons within that Vendor's organization who will be directly involved with the project in the various capacities.
3. Confirm the names, addresses, and phone numbers of all Sub-Subcontractors and Sub-Suppliers. This information will:

    *a.* Correlate the presence of Sub-Vendors with those indicated in the subcontract and purchase order process.
    *b.* Identify those Vendors who may be considered for joint payments.
    *c.* Allow initiation of contact directly with the Sub-Vendors, which will greatly facilitate later material expediting efforts (see Section 4.11).

4. Update the Company file on the respective Subcontractor or Supplier.
5. Determine or confirm the relative and absolute amounts of work that the Sub-Vendor is actually performing with its own forces and those that are being sub-sublet.

    *a.* The identification of all such sublet sources early on (before they become "issues") makes confirmation of all material scope, delivery, and cost information directly with the sources a matter of simple phone calls.

### 4.5.3 Procedure and Distribution

1. Send the reference form together with the subcontract or purchase order attached to the *Subcontract/Purchase Order Transmittal Form Letter* of Section 4.7.
2. Enforce its return *with the subcontract or purchase order.*
3. Upon receipt, immediately contact the Vendor to complete any relevant information that is apparently missing.

    *a.* The most obvious omission will likely be the home and car phone numbers of key production people. The nature of this business is that if they're deserving of your order, you're deserving of the ability to contact these individuals in emergency or expediting situations. Do whatever is necessary to get these numbers *now*—well before you will need them.

4. Distribute the *completed* form to:

    *a.* The Sub-Vendor's general file
    *b.* The Subcontractor Summary Log of Section 4.4
    *c.* The Site Superintendent

### 4.5.4 Sample Subcontractor/Supplier Reference Form *(page 4.14)*

The Sample Subcontractor/Supplier Reference Form that follows is self-explanatory. It is to be used to follow through on the procedure described in this section.

## 4.5.4
## Sample Subcontractor / Supplier Reference Form

PROJECT________________________ SECTION(S) ________________
COMPANY No.________________________ ________________

Please complete all applicable items below and return this form in order to facilitate communication between our organizations. Should you desire similar information from our company, please do not hesitate to ask.

Your correct mailing address is ________________________
(Include street location) ________________________
________________________
________________________

Phone Number: _____-_____-_____ Fax Number: _____-_____-_____

PRESIDENT________________________
VICE PRESIDENT________________________

Person responsible for PROJECT COORDINATION:
Name:________________________
Title:________________________
Phone: Business ____________
Car ____________
Home ____________
Fax ____________

Person responsible for APPROVAL SUBMISSIONS:
Name:________________________
Title:________________________
Phone: Business ____________
Car ____________
Home ____________
Fax ____________

FIELD SUPERINTENDENT:
Name:________________________
Title:________________________
Phone: Business ____________
Car ____________
Home ____________
Fax ____________

PRODUCTION MANAGER:
Name:________________________
Title:________________________
Phone: Business ____________
Car ____________
Home ____________
Fax ____________

Person responsible for PAYMENT APPLICATIONS:
Name:________________________
Title:________________________
Phone: Business ____________
Car ____________
Home ____________
Fax ____________

PRESIDENT________________________
CREDIT MANAGER________________________

Your Subcontractors and Material Suppliers are:

ITEM:________________________
Company________________________
Street________________________
City____________State________Zip________
Phone____________Fax____________
Contact________________________

ITEM:________________________
Company________________________
Street________________________
City____________State________Zip________
Phone____________Fax____________
Contact________________________

ITEM:________________________
Company________________________
Street________________________
City____________State________Zip________
Phone____________Fax____________
Contact________________________

ITEM:________________________
Company________________________
Street________________________
City____________State________Zip________
Phone____________Fax____________
Contact________________________

ITEM:________________________
Company________________________
Street________________________
City____________State________Zip________
Phone____________Fax____________
Contact________________________

A master copy of the form can be prepared that includes the project references and any supplementary information specific to the project which is not included on the standard form. Photocopy a supply of the project forms for use.

## 4.6 Transmittal Form Letter Procedure and Use

### 4.6.1 General

The Transmittal Form Letter is one of the most common types of correspondence used throughout the industry. Its purpose is to document the movement of information and/or materials. Each time a shop drawing is transferred from the custody of one party to that of another, for example, it should be done by way of a transmittal.

Most individuals, however, do not use the form properly. In the rush to fill it out and get it over with:

- Individual names are neglected (personal accountability).
- File references get omitted.
- Reasons for the transmittal or required action are not clearly indicated.
- The method of delivery is left out.
- Important remarks are abbreviated or otherwise left unclear.
- Distribution of the item or of the transmittal information itself is not properly performed.

### 4.6.2 Proper Use of the Transmittal Form Letter

Do not let your transmittal become a low-value, time-wasting exercise. Follow these few rules as a routine application that will leave complete and comprehensive records where and when you need them. Use the Transmittal Form Letter when sending all shop drawings, samples, copies of letters, or almost anything else. In each case, be sure to:

1. *Send it to a specific person's attention.* If you don't know who will be handling the response, find out.
2. *Always note the complete file reference.* Refer to Section 2.3, *Correspondence,* and Section 2.4, *Files and File Management.* Indicate the designated file instruction on the Transmittal Form Letter in the same way as on any other correspondence.
3. *Indicate definite action on the letter.* The boxes on the form are there to be checked. Use them. If a document is attached, note it. If it has been sent under separate cover, indicate it properly as such.
4. *Indicate the method of delivery.* This can be a very important piece of information, particularly in cases where time or delay in response becomes an issue. It might make all the difference in the world, for example, whether the document had been faxed, mailed, sent express delivery, carrier pigeon, or had been hand delivered.
5. *Use document identification numbers if there are any.* Describe the item(s) being sent concisely, *but in a manner that identifies them positively and distinguishes them from other items.* A description of "Duct Drawings," for example, means

absolutely nothing. Use instead the same description that is on the drawing's title block.

6. *Request definite action.* Again, the boxes are there to be checked off. Do not assume that the architect "knows" that you are expecting his or her office to "approve" the submission. If you need a response that has not been conveniently categorized for you, don't settle for what's there. Take the time to be certain that you indicate positively what action you need.
7. *Use the "remarks" area for short but important qualifications and notifications.* This is a legitimate project record document. The recipients are responsible for all information contained in it. Don't overlook opportunities for repeated action requests and other notifications.
8. *Indicate the correct distribution of the letter and its attachments.* Refer to Section 2.3.3, *Correspondence Distribution,* for specific instructions in this regard.

### 4.6.3 Sample Letter of Transmittal—Completed Example *(page 4.17)*

A completed example of the Sample Letter of Transmittal is provided here to display the approach presented in this section. Section 4.6.4 provides a blank form to be copied for actual use.

### 4.6.4 Sample Letter of Transmittal — Blank Form *(page 4.18)*

## 4.7 Subcontract and Purchase Order Distribution Procedure

### 4.7.1 General

The descriptions of the subcontract and purchase orders themselves, along with a discussion of relevant clauses and their applications, are specifically developed elsewhere.

This section describes the execution and distribution procedure to be followed, after a complete agreement has been reached, that will:

- Expedite the process
- Preclude the possibility of unauthorized modifications
- Distribute the documents to all those who need them in order to properly administer the project
- Preserve the security of originally executed documents

### 4.7.2 Procedure

1. Five copies of the subcontract or purchase order are prepared.
2. The *Subcontract/Purchase Order Transmittal Form Letter* of Section 4.7.3 is completed as appropriate.

   *a.* The first item is checked noting that (three) *unsigned* copies of the subcontract or purchase order are attached, and indicates that *all three* are to be executed and returned to the Company for final execution. This may appear

## 4.6.3 Sample Letter of Transmittal (Completed Example #1)

### LETTER OF TRANSMITTAL

To: STRAIT STEEL CORP.
456 SUMMER ST.
STAMFORD, CT 06362
Attn: JOHN STRAIT

Date: OCT 1, 1994
Project: BRIDGETOWN TREATMENT PLANT
Project #: 9660
Subject: ANCHOR BOLTS

TRANSMITTED: X Attached ___ Under Separate Cover Via: X Mail ___ Express Mail ___ Fax: # ______
are the following:

X Shop Drawing(s) ___ Contract Drawing(s) ___ Specification(s)
___ Letter(s) ___ Coordination Drawing(s) ___ Change Order(s)
___ Sample(s) ___ Message (See Remarks) ___ ______

| Quant | Date | ID No. | Description |
|---|---|---|---|
| 2 | 9.19.94 | AB-1 | ANCHOR BOLT LAYOUT & DETAILS |
| | | | |
| | | | |
| | | | |
| | | | |
| | | | |
| | | | |
| | | | |
| | | | |
| | | | |
| | | | |

**If items are not received as listed, please notify us immediately.**

___ For Approval ___ For Your Use ___ As Requested ___ For Fabrication
___ Approved X Approved As Noted ___ Rejected ___ Revise & Resubmit

Remarks: PLEASE CONFIRM BY 10.5.94 THAT ALL ANCHOR BOLTS & CAGES WILL BE ON-SITE BY 10.15.94

cc: JOBSITE W/ ATT.
BENT STEEL ERECTORS W/ATT.

File: STRUCT. STEEL W/ATT

CF

COMPANY

David Jeffko
D. JEFFKO, PROJECT ENGINEER

## 4.6.4
## Sample Letter of Transmittal

### LETTER OF TRANSMITTAL

To: ______________________
______________________
______________________

Attn: ______________________

Date:______________________
Project: ______________________
Project #:______________________
Subject:______________________

TRANSMITTED: ___ Attached ___Under Separate Cover Via: ___Mail ___Express Mail ___Fax: #___________
are the following:
___ Shop Drawing(s) ___ Contract Drawing(s) ___ Specification(s)
___ Letter(s) ___ Coordination Drawing(s) ___ Change Order(s)
___ Sample(s) ___ Message (See Remarks) ___ ______________________

| Quant | Date | ID No. | Description |
|---|---|---|---|
| | | | |
| | | | |
| | | | |
| | | | |
| | | | |
| | | | |
| | | | |
| | | | |
| | | | |
| | | | |
| | | | |
| | | | |

**If items are not received as listed, please notify us immediately.**

____ For Approval ____ For Your Use ____ As Requested ____ For Fabrication
____ Approved ____ Approved As Noted ____ Rejected ____ Revise & Resubmit
____ __________ ____ __________ ____ __________ ____ __________

Remarks:
______________________________________________
______________________________________________
______________________________________________
______________________________________________
______________________________________________
______________________________________________
______________________________________________
______________________________________________
______________________________________________
______________________________________________
______________________________________________
______________________________________________
______________________________________________

cc: ______________________
______________________
______________________

File: ______________________
______________________
CF

COMPANY

______________________
______________________

to be a simple detail, but in practice it will stop any unauthorized attempts to modify the agreement in any way. If, for example, the documents were sent out *with* signatures, there would be no control over anyone wishing to modify the signed document. While it is true that it may be argued later on technical grounds that provisions so modified needed to be initialed by both parties, it sets the stage for significant arguments and problems at the jobsite.

*b.* Other beginning requirements, such as delivery of the certificate of insurance and of the proper submittals, are also highlighted.

3. Copies of the Transmittal Form Letter and the subcontract or purchase order documents are distributed to:

   *a.* The respective Vendor project file (to maintain copies of the agreement as actually tendered)
   *b.* The Correspondence File (CF) as backup

4. The three copies of the executed subcontract or purchase order and the complete Certificate of Insurance *in proper form and coverage amounts* must then be *delivered* by the Subcontractor or Supplier to the Company *before any work is allowed to proceed at the site.* Under *no* circumstances will exception to this requirement be allowed.
5. If the subcontract or purchase order forms have been modified in any way when they are returned, the modifications must be dealt with immediately in a manner consistent with the specific contract approval authorities of the Company.
6. The three copies of the final documents are executed by the Company. The distribution will be made as follows:

   *a.* One original is returned to the Vendor using another *Subcontract/ Purchase Order Transmittal Form Letter* of Section 4.7.3, completing the lower portion of the form letter as appropriate.
   *b.* One original is distributed to the fireproof vault at Accounts Payable that houses all contracts.
   *c.* One original is distributed to the respective Vendor file in the *General Project File* as set up per direction of Section 2.4.4, with duplicate copy to the Correspondence File.

7. One additional copy of the subcontract or purchase order is made deleting all prices. This modified copy will be distributed to the jobsite for use in confirming the actual scope of the Vendor's responsibilities.

### 4.7.3 Sample Subcontract/Purchase Order Transmittal Form Letter—Completed Example *(page 4.20)*

The Sample Subcontract/Purchase Order Transmittal Form Letter included here is self-explanatory. It will be used as described in Section 4.7.2 to send copies of unexecuted documents for execution, copies of executed documents, and notification of other requirements. Check off the appropriate box for the distribution process being performed.

### 4.7.4 Sample Subcontract/Purchase Order Transmittal Form Letter—Blank Form *(page 4.21)*

## 4.7.3
## Sample Subcontract/P.O. Transmittal Form Letter
## (Completed Example)

| Letterhead |
|---|

To: TRIANGLE MASONRY CORP.
555 TERMINAL RD.
BRITTON, CT 06555

Date: 3·28·94

ATTN: JOHN TRIANGLE

RE: Project: ENGINE CO. #4
Company Project #: 9440

SUBJ: Subcontract (Purchase Order) Agreement

Mr. ~~(Ms.)~~ TRIANGLE:

Please find attached:

X Three copies of the Subcontract Agreement ~~(Purchase Order)~~. Please sign and return *all three copies* for final execution.
___ One executed copy of the Subcontract Agreement (Purchase Order). Retain for your records.
X The Subcontract/Supplier Reference Form. Complete as appropriate and return to my attention.
___ Three copies of the Full Lien Waiver Forms to be executed and returned with the Subcontract or Purchase Order Agreement.
X ADD'L COPY OF COMPLETE PLANS & SPECS
___ ______

Please:

X Deliver all appropriate Certificated of Insurance in correct coverage amounts. Be sure to include *all* requirements for additional insured.
X Deliver 9 copies of all required approval and coordination submittals to:
X This Office ___ The Jobsite
X Confirm that your current delivery and erection schedules meet current Project Schedule Requirements as coordinated with ANTHONY BARTA, Project Superintendent.
___ ______
___ ______

Very truly yours,

COMPANY

George Raymond

GEORGE RAYMOND
Project Manager

cc: A/P, Jobsite
File: Subcontracts, CF
Sub-Vendor File: MASONRY

## 4.7.4
## Sample Subcontract/P.O. Transmittal Form Letter

| Letterhead |
|---|

To: ______________________ Date: ______________________
______________________
______________________

ATTN: ______________________

RE: Project: ______________________
Company Project #: ______________________

SUBJ: Subcontract (Purchase Order) Agreement

Mr. (Ms.) ______________________:

Please find attached:

___ Three copies of the Subcontract Agreement (Purchase Order). Please sign and return *all three copies* for final execution.
___ One executed copy of the Subcontract Agreement (Purchase Order). Retain for your records.
___ The Subcontract/Supplier Reference Form. Complete as appropriate and return to my attention.
___ Three copies of the Full Lien Waiver Forms to be executed and returned with the Subcontract or Purchase Order Agreement.
___ ______________________
___ ______________________

Please:

___ Deliver all appropriate Certificated of Insurance in correct coverage amounts. Be sure to include *all* requirements for additional insured.
___ Deliver _____ copies of all required approval and coordination submittals to:
___ This Office ___ The Jobsite
___ Confirm that your current delivery and erection schedules meet current Project Schedule Requirements as coordinated with ______________________, Project Superintendent.
___ ______________________
___ ______________________

Very truly yours,

COMPANY

Project Manager

cc: A/P, Jobsite
File: Subcontracts, CF
Sub-Vendor File: __________

### 4.7.5 Subcontract/Purchase Order Distribution Flowchart

The flowchart displayed here presents the document flow process described in Section 4.7.2 for clarification.

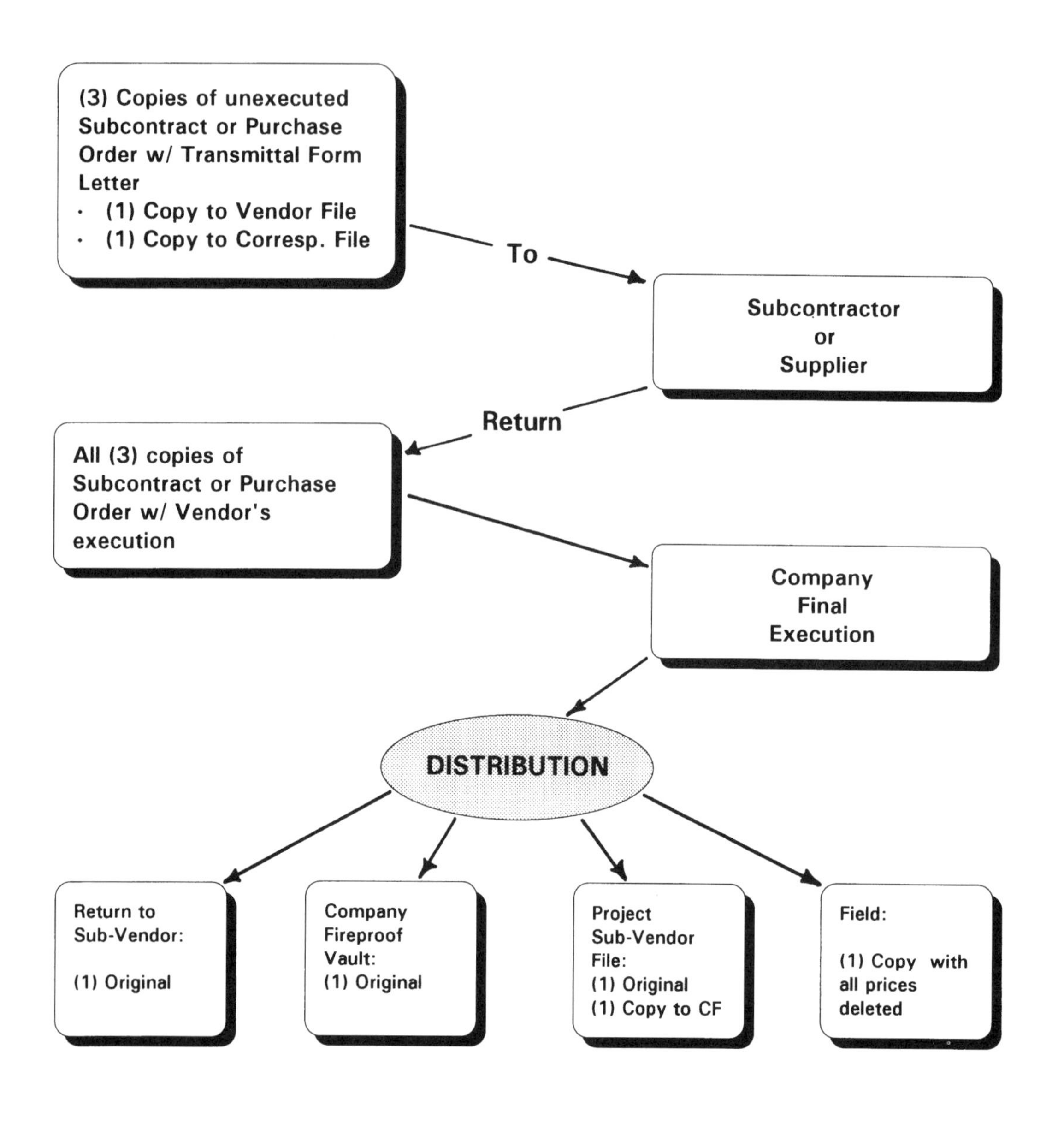

## 4.8 Subcontractor Schedule of Values

### 4.8.1 General

There are a number of requirements that affect the Subcontractor's proper Schedule of Values materially, including:

- Time of submission
- Level of detail
- Authority to approve or reject its composition
- Amount of work performed by the Subcontractor's own forces
- Amount of work performed by Sub-Subcontractors
- Value of materials provided
- Value, source, and composition of equipment provided
- Value of final tie-in and system operation activities
- Value of all project close-out efforts
- Value of guarantees, warranties, instruction, manuals, etc.

These considerations necessarily vary greatly between bid packages for both practical and firm contractual reasons. A framing contractor, for example, will not have the same "final tie-in" or "close-out efforts" as the HVAC contractor. Its level of necessary detail will also be much less. That same framing contractor, for example, is not likely to have long-distance suppliers of expensive, sensitive equipment that has elaborate guarantee requirements.

The requirements of each bid package must therefore be considered individually and have *all* considerations applied to it in order to confirm the final determination of all its particular requirements.

Refer to Section 3.9, *Schedule of Values,* and Section 3.10, *Requisitions for Payment and Contract Retainage,* for important related discussion.

### 4.8.2 Time of Submission

It is crucial that the Subcontractor's Schedule of Values be delivered to the Company in the absolute shortest period of time. The general Schedule of Values of the Company to the Owner is one of the earliest documents to be expedited if the first payments are to be processed to all Vendors on time. Further, every Sub-Vendor's schedule must correlate directly with the manner in which its prices have been represented to the Owner, or there will be major scope/value payment disputes between the Company and its Sub-Vendors in every payment cycle. These considerations cannot be stressed enough.

Refer to Sections 4.7 and 4.8 regarding subcontracts. There you will find instructions for indicating the Sub-Vendor's Schedule of Values directly on the subcontract or purchase order form itself for those more simply defined bid packages.

Use the principles of this section and of Sections 3.9 and 3.10 to properly determine that Sub-Vendor's schedule, and whenever the opportunity is available to do so, have the entire matter put to rest at the time of subcontract or purchase order execution.

If the timing of the bid package or other considerations do not allow for the immediate finalization of the completed, detailed Sub-Vendor's Schedule of Values, follow the directions of the later portions of this section to expedite the complete process.

### 4.8.3 Correlation with the General Schedule of Values

Section 3.10.3 discusses this consideration from the General Schedule of Values perspective. The most common dilemma is that often the General Schedule of Values must be submitted to the Owner long before the Schedules of Values of major or complex bid packages are finalized between the General Contractor and the respective Sub-Vendors. An approach that can help to deal with the problem, thereby avoiding continuing payment issues with both the Owner and the Sub-Vendors, follows:

1. Prepare the General Schedule of Values to the level of available detail with respect to the work of all technical bid packages.
2. For any bid package that will be submitting requisitions for work in the first requisition period (such as site work or concrete) there is no option. Those bid package Schedules of Values *must* be completely resolved between the Company and the respective Subcontractors by the time the General Schedule of Values is submitted.
3. Many (or at least several) bid packages for work that will not be performed during the first payment period may not yet have had their detailed breakdowns completely resolved as of the time of the first submission of the General Schedule of Values.

   For these cases include a lump sum for the entire bid package. This will allow the Owner to approve the relative amount of the bid package in the approval of the General Schedule, while it will allow you flexibility to provide ultimately a breakdown and level of detail that directly correlate to the Sub-Vendor's Schedule of Values for its billing throughout the remainder of the project.

   Review this procedure completely with the Owner and design professionals before proceeding. Note in those discussions that billings against such lump sums will not occur until the appropriate breakdowns have been submitted and approved. Having the total approach properly explained is usually sufficient to confirm the procedure.
4. When the Sub-Vendor's Schedule of Values is finalized in accordance with the recommendations of this section, immediately submit it for approval by the Owner in the same procedure as the original General Schedule of Values. Upon approval, include the additional detail on the General Schedule for use as such throughout the remainder of the project.

### 4.8.4 Level of Detail

Section 3.9.3 describes these considerations from the perspective of the General Schedule of Values. Here note first that most Subcontractors have a tendency to resist providing sufficient levels of detail in their Schedules of Values until forced to do so.

The procedure to determine the appropriate level of detail is very straightforward. In its simplest idea, it boils down to the following:

1. Visualize the manner and sequence in which the work will be accomplished physically.

2. Visualize the determination that will become necessary in order to demonstrate the level of completion of a particular component, and to provide some facility to the Owner's representatives to verify the bill.

   *a.* Is a single item titled *Branch Wiring* sufficient, for example, to determine the actual amount of work completed without going through financial gymnastics each period, or will it be necessary to detail the individual levels of completeness in various areas of the project?
   *b.* Will *cubic feet of concrete* be enough to properly determine the correct relative amount of reinforcing steel throughout the footings, walls, columns, beams, and so on, or will you wind up in a "discussion" every time the Owner will want to release payment for steel on a prorated basis while your actual payables to the steel vendor are heavy in a certain area?

3. Separate Material and Labor Prices

   *a.* Do this first as a practical matter in helping to improve the detail that will facilitate the other considerations of this section.
   *b.* Second, it may turn out to be one of the "unwritten rules" of the Owner to pay only for materials after they have been installed if there is no approved breakdown to the contrary.

      This practice can have the potential for disaster if, for example, you or your Subcontractor is on a 30-day payment cycle and the material is delivered on the 5th of the month, but won't be installed until the 20th of the following month.

      Directly avoid the problem by having all material identified separately, billable per the express terms of the Owner/Contractor agreement.

4. Determine and/or confirm appropriate unit price values as a method of at least generally confirming the relative correctness of the billing items.
5. Give careful and detailed attention to all project close-out activities. Do not overlook the values of such items as:

   *a.* System tie-ins and balancing
   *b.* Hookups or other connection with work of other trades
   *c.* Submission of guarantees, warranties, maintenance and operating manuals, as-built drawings, and other special documentation
   *d.* Special instruction to be provided to Owner maintenance personnel
   *e.* "Incidentals," such as labeling, key plans and charts, and color-coding of components

6. Give special attention to confirming the complete, real values of the remaining 25 to 30 percent of the work.

   *a.* If left to its own process, it will be a sure bet that by the time the work has been approved to be "75 percent complete," there would not be a Contractor on the planet—including the one who is responsible—who could actually complete the work for the remaining 25 percent of money available.

      In the worst of these situations the Sub-Vendor can leave or be terminated at a late point of completion, leaving everyone with the major problem of completing the work for the lesser amount of available money remaining. In the best situations it will be like pulling teeth to have that Sub-Vendor who stays with the job finish its odds and ends on time, complete its punchlist, and provide its documentation. *Keep appropriate values assigned to all items, and the work stands a much greater chance of getting done without major incident.*

## 4.8.5 Sample Sub-Vendor Schedule of Values/Requisition for Payment Form

The Sample Sub-Vendor Schedule of Values/Requisition for Payment Form included here is to be sent to each Vendor for their use in properly representing their Schedules of Values and subsequent applications for payment.

In cases where the Schedule of Values has been finalized by subcontract execution, the line-item listing will correlate directly with that indicated in the subcontract or purchase order. In cases where the Sub-Vendor's Schedule of Values is still to be submitted and approved, it will become the form of submission.

In either case, once the Sub-Vendor's Schedule of Values is finally approved, it will become the periodic requisition form for that bid package for the life of the project. Extras, credits, and backcharges will be added as new line items as the project progresses.

#### 4.8.5.1 Sample Sub-Vendor Schedule of Values/Requisition For Payment Form—Completed Example #1 *(page 4.27)*

#### 4.8.5.2 Sample Sub-Vendor Schedule of Values/Requisition for Payment Form—Completed Example #2 *(page 4.28)*

#### 4.8.5.3 Sample Sub-Vendor Schedule of Values/Requisition for Payment Form—Completed Example #3 *(page 4.29)*

#### 4.8.5.4 Sample Sub-Vendor Schedule of Values/Requisition for Payment Form—Blank Form *(page 4.30)*

## 4.8.6 Sample Sub-Vendor Schedule of Values/Requisition for Payment Form Letter *(page 4.31)*

The Sample Sub-Vendor Schedule of Values/Requisition for Payment Form Letter that follows is used to:

1. Transmit the Schedule of Values/Requisition for Payment Forms
2. Advise of the manner of preparation of the Sub-Vendor's Schedule of Values
3. Notify of the required submission date
4. Identify the approved form as the future requisition form
5. "Remind" the Sub-Vendor of its contractual requirement to submit its periodic requisition on time in order to avoid problems

Use this procedure to expedite all Sub-Vendor Schedules of Values that, for whatever reason, have not been able to be finalized as of the contract execution date.

## 4.8.7 Sample Sub-Vendor Schedule of Values Approval/Rejection Form Letter *(page 4.32)*

The Sample Schedule of Values Approval/Rejection Form Letter that follows is used to return the Sub-Vendor's Schedule of Values submission, directly indicating that either:

# 4.8.5.1
# Sample Schedule of Values/Requisition for Payment Form (Completed Example #1)

## SCHEDULE OF VALUES / REQUISITION FOR PAYMENT

Co: HEAT & AC, INC.
Proj.: FIREHOUSE ADD'N
No.: 9424
Period:

Application No.:
Date:

Bid Package: HVAC

Page 1 of 1

| SCHEDULE OF VALUES | | | | | REQUISITION FOR PAYMENT | | | | |
|---|---|---|---|---|---|---|---|---|---|
| Item # | Description | Quant | Unit Price | Total | Prev. Billed | This Application | Total Complete To Date | % Comp | Balance To Complete |
| 1 | SUBMITTALS | L.S. | L.S. | 2,000.- | | | | | |
| 2. | STL. DUCT MAINS | 230 LF | 20.- | 4,600.- | | | | | |
| 3. | STL. DUCT BRANCHES | 570 LF | 15.- | 8,550.- | | | | | |
| 4. | DUCT INSULATION | 800 LF | 4.- | 3,200.- | | | | | |
| 5. | FAN-COIL UNITS (MAT) | 10 UN | 900.- | 9,000.- | | | | | |
| 6. | FAN-COIL UNITS (LAB) | 10 UN | 300.- | 3,000.- | | | | | |
| 7. | BOILER (MAT) | 1 UN | 3,500.- | 3,500.- | | | | | |
| 8. | BOILER (LAB) | 1 UN | L.S. | 2,000.- | | | | | |
| 9. | CHILLER (MAT) | 1 UN | 3,000.- | 3,000.- | | | | | |
| 10. | CHILLER (LAB) | 1 UN | L.S. | 2,000.- | | | | | |
| 11. | TEMP. CONTROLS | L.S. | – | 2,600.- | | | | | |
| 12. | SYSTEM BALANCING | L.S. | – | 3,200.- | | | | | |
| 13. | AS-BUILTS / O&M MAN. | L.S. | – | 2,500.- | | | | | |
| 14. | GUARANTIES / WARRANT. | L.S. | – | 2,000.- | | | | | |
| | TOTALS | | | $51,150.- | | | | | |

## 4.8.5.2
## Sample Schedule of Values/Requisition for Payment Form (Completed Example #2)

**SCHEDULE OF VALUES / REQUISITION FOR PAYMENT**

Co: HEAT & AC, INC. Application No.: 1 Bid Package: HVAC
Proj.: FIREHOUSE ADD'N Date: APRIL 30, 1994
No.: 9424
Period: APRIL, 1994 Page 1 of 1

| SCHEDULE OF VALUES | | | | | REQUISITION FOR PAYMENT | | | | |
|---|---|---|---|---|---|---|---|---|---|
| Item # | Description | Quant | Unit Price | Total | Prev. Billed | This Application | Total Complete To Date | % Comp | Balance To Complete |
| 1. | SUBMITTALS | L.S. | – | 2,000.- | | 2,000.- | 2,000.- | 100 | 0.0 |
| 2. | STL. DUCT MAINS | 230 LF | 20.- | 4,600.- | | 3,680.- | 3,680.- | 80 | 920.- |
| 3. | STL. DUCT BRANCHES | 570 LF | 15.- | 8,550.- | | 4,275.- | 4,275 | 50 | 4,275.- |
| 4. | DUCT INSULATION | 800 LF | 4.- | 3,200.- | | | | 0 | 3,200.- |
| 5. | FAN-COIL UNITS <MAT> | 10 UN | 900.- | 9,000.- | | | | 0 | 9,000.- |
| 6. | FAN-COIL UNITS <LAB> | 10 UN | 300.- | 3,000.- | | | | 0 | 3,000.- |
| 7. | BOILER <MAT> | 1 UN | 3,500.- | 3,500.- | | | | 0 | 3,500.- |
| 8. | BOILER <LAB> | 1 UN | L.S. | 2,000.- | | | | 0 | 2,000.- |
| 9. | CHILLER <MAT> | 1 UN | 3,000.- | 3,000.- | | | | 0 | 3,000.- |
| 10. | CHILLER <LAB> | 1 UN | L.S.- | 2,000.- | | | | 0 | 2,000.- |
| 11. | TEMP. CONTROLS | L.S. | – | 2,600.- | | | | 0 | 2,600.- |
| 12. | SYSTEM BALANCING | L.S. | – | 3,200.- | | | | 0 | 3,200.- |
| 13. | AS-BUILTS/O&M MAN. | L.S. | – | 2,500.- | | | | 0 | 2,500.- |
| 14. | GUARANTIES/WARRANT. | L.S. | – | 2,000.- | | | | 0 | 2,000.- |
| TOTALS | | | | $51,150.- | | 9,955.- | 9,955 | 19.46 | 41,195.- |

## 4.8.5.3 Sample Schedule of Values/requisition for Payment Form (Completed Example #3)

**SCHEDULE OF VALUES / REQUISITION FOR PAYMENT**

Co: HEAT & AC, INC. Application No.: 2 Bid Package: HVAC
Proj.: FIREHOUSE ADDITION Date: MAY 31, 1994
No.: 9242
Period: MAY 1994

Page 1 of 1

| SCHEDULE OF VALUES | | | | | REQUISITION FOR PAYMENT | | | | |
|---|---|---|---|---|---|---|---|---|---|
| Item # | Description | Quant | Unit Price | Total | Prev. Billed | This Application | Total Complete To Date | % Comp | Balance To Complete |
| 1. | SUBMITTALS | L.S. | - | 2,000.- | 2,000.- | 0.- | 2,000.- | 100 | 0.- |
| 2. | STL. DUCT MAINS | 230 LF | 20.- | 4,600.- | 3,680.- | 920.- | 4,600.- | 100 | 0.- |
| 3. | STL. DUCT BRANCHES | 570 LF | 15.- | 8,550.- | 4,275.- | 3,420.- | 7,695.- | 90 | 855.- |
| 4. | DUCT INSULATION | 800 LF | 4.- | 3,200.- | | 1,600.- | 1,600.- | 50 | 1,600.- |
| 5. | FAN COIL UNITS (M) | 10 UN | 900.- | 9,000.- | | | | 0 | 9,000.- |
| 6. | FAN COIL UNITS (L) | 10 UN | 300.- | 3,000.- | | | | 0 | 3,000.- |
| 7. | BOILER (MAT) | 1 UN | 3,500.- | 3,500.- | | 3,500.- | 3,500.- | 100 | 0.- |
| 8. | BOILER (LAB) | 1 UN | L.S. | 2,000.- | | 1,000.- | 1,000.- | 50 | 1,000.- |
| 9. | CHILLER (MAT) | 1 UN | 3,000.- | 3,000.- | | 3,000.- | 3,000.- | 100 | 0.- |
| 10. | CHILLER (MAT) | 1 UN | L.S. | 2,000.- | | | | 0 | 2,000.- |
| 11. | TEMP. CONTROLS | L.S. | - | 2,600.- | | | | 0 | 2,600.- |
| 12. | SYSTEM BALANCING | L.S. | - | 3,200.- | | | | 0 | 3,200.- |
| 13. | AS-BUILTS / O & M MAN. | L.S. | - | 2,500.- | | | | 0 | 2,500.- |
| 14. | GUARANTIES/WARRANT. | L.S. | - | 2,000.- | | | | 0 | 2,000.- |
| 15. | CO#1 - CHANGES TO FAN-COIL UNITS | - | - | 3,000.- | | | | 0 | 3,000.- |
| 16. | CO#2 - DUCT CHANGES | - | - | 2,000.- | | 1,800.- | 1,800.- | 90 | 200.- |
| **TOTALS** | | | | 56,150.- | 9,955.- | 15,290.- | 25,195.- | 44.87 | 30,955.- |

# 4.8.5.4
# Sample Schedule of Values/Requisition for Payment Form

**SCHEDULE OF VALUES / REQUISITION FOR PAYMENT**

Co:
Proj.:
No.:
Period:

Application No.:
Date:

Bid Package:

Page of

| SCHEDULE OF VALUES | | | | | REQUISITION FOR PAYMENT | | | | |
|---|---|---|---|---|---|---|---|---|---|
| Item # | Description | Quant | Unit Price | Total | Prev. Billed | This Application | Total Complete To Date | % Comp | Balance To Complete |
| | | | | | | | | | |
| | TOTALS | | | | | | | | |

## 4.8.6
## Sample Schedule of Values/Requisition for Payment Instruction Form Letter

**Letterhead**

Date: ______________________

To : _________________________________
_________________________________
_________________________________

ATTN: _________________________________

RE: Project:________________________
Project #:______________________

SUBJ: Schedule of Values/Requisition for Payment

Mr. (Ms.)__________________________:

Attached are the Schedule of Values/Requisition for Payment Forms, to be used as follows:

1. Prepare & Submit your Schedule of Values

   On the left side of the form, separate your total contract into individual line items that accurately reflect the manner in which material and equipment will be delivered to the jobsite, and the way in which the work will be performed. Provide sufficient level of detail that will allow the Owner, Architect, and this office to properly evaluate your periodic requisitions for payment. Also include appropriate categoies of your general conditions and project close-out items.

   Submit your Schedule of Values to this office by ____________________, 19_____ for approval.

2. Requisition for Payment

   When the Schedule of Values is approved, it becomes your Requisition for Payment Form.

   By the 25th of each month, submit your Requisition on the same form by indicating on the right side of the form the amounts completed for each line item. Compliance with this procedure is mandatory if your payments are to be processed for the respective period.

Please call if you have any questions.

Very truly yours,

COMPANY

_______________________________________
Project Manager

cc: File:____________, CF
A/P

## 4.8.7
## Sample Sub-Vendor Schedule of Values Approval/Rejection Form Letter

| Letterhead |
|---|

Date: ______________________________

To: ______________________________
______________________________
______________________________

ATTN: ______________________________

RE: Project:______________________________
Company Project #: ______________________

SUBJ: Schedule of Values

Mr. (Ms.) ______________________________:

Attached is:

___ Your Schedule of Values approved as submitted. Please use this form on all periodic Requisitions for Payment that are to be submitted to this office by the 25th of each month.

___ Your Schedule of Values marked to suggest more appropriate breakdowns, and/or with questions regarding the submission. Please revise and resubmit as soon as possible in order to avoid delay in processing your requisitions

Please call if you have any questions.

Very truly yours,

COMPANY

______________________________
Project Manager

cc: A/P w/att.
File: Sub-Vendor File:________
Subcontracts, CF

- It is approved as it is, and is accordingly to be used as the basis for all applications for payment, or
- It is not approved, along with an appropriate modification that will allow its approval.

Use it to expedite processing of problem submissions, and to confirm the correct paper trail and chronology for those cases where payment or payment timing to Sub-Vendors is affected.

## 4.9 Shop Drawing and Submittal Management

### 4.9.1 Operating Objectives of the Project Engineer

Shop drawings and other submittals for approval are required in virtually every bid package. It is the responsibility of the Project Engineer to:

1. Identify all submittals required to be processed for each bid package
2. Coordinate the schedule of submittals with the Purchase and Award Schedule in order to ensure that all submittal and review processes are conducted within the time requirements of the progress schedule
3. Police the performance of each Sub-Vendor to enforce compliance with all document requirements with respect to form, content, and time
4. Review each Sub-Vendor's submittals for compliance with the contract documents and with its own subcontract or purchase order, considering scope of work and coordination with contiguous work
5. Confirm that each submittal is in the proper form and includes the correct number of copies as required for all submission and distribution requirements with respect to both specified items and any special requirements for the submission of anything to be considered an equal or a substitution (refer to Section 3.7 for important related discussion)
6. Keep all submittals moving smoothly through each designer's office to ensure that everyone takes appropriate action within the "reasonable period of time" required in the specifications
7. Keep the design professionals' actions appropriate
8. Be prepared to take quick, decisive action in the face of incorrect, late, or otherwise inappropriate submittal review action on the part of the Architect and Engineers
9. Distribute copies of all submittals and other appropriate information to everyone who will need them—in good time as dictated by the progress schedule

For important directly related discussion refer to:

- Section 2.4, *Files and File Management*
- Section 3.3.6, *"Performance" and "Procedure" Specifications*
- Section 3.3.17, *Proprietary Specifications*
- Section 3.6, *Shop Drawing "Approval"*
- Section 3.7, *Equals and Substitutions*

### 4.9.2 Submittal Log

The Submittal Log is a collection of forms to be used on each project to catalog each document's flow through the entire approval and distribution process. It is the place in which to record each submittal's respective characteristics along with the resulting action by each party and all distribution information.

It provides at a glance:

- The listing of all submittals as required (This will serve as an important checklist to prevent overlooking any requirement.)
- The chronological order of submissions
- The required and actual dates of each submission
- The date by which designer action is needed
- The actual Owner or designer action and dates
- The required and actual distribution

The log will be an important document control tool that will be key to expediting the complete and comprehensive submittal control effort.

### 4.9.3 Submittal Log Procedure

1. Place a supply of forms in a 1-inch three-ring loose leaf binder. Depending upon the size of the project, it may be reasonable to combine this log with the *Subcontractor Summary* and the *Telephone Log* of Section 4.4 in the same binder.
2. Arrange the forms by specification section or by bid package number, identified in the same manner as the *General Project File* described in Section 2.4.4.
   *a.* This correlates the Submittal Log directly with the complete project file.
   *b.* It provides a research method based on the specification as a second way to research subcontractor information (the first being the alphabetical method of the Subcontractor Summary).
3. Research each individual specification section for each bid package. List all specific submissions required on the respective Vendor's log form.
4. Determine any other submittal not yet specifically requested that you may require for proper coordination of the work. Add to each Vendor's list as the shop drawing approval process for all Subcontractors proceeds. The process itself will often identify such necessary or desired submittals.
5. Insert the name of each Subcontractor who will be requiring a copy of the approved submittal for coordination in one of the headings "______ COPY."
   *a.* For example, if you know (or have been told) that the Concrete Subcontractor will need the Steel Subcontractor's anchor bolt layout, insert the name of the Concrete Subcontractor in one of the columns in the Steel Subcontractor's log form. Place a check mark in the corner of the Concrete Subcontractor's distribution box adjacent to the anchor bolt submittal line. This will remind you of the distribution requirement when you are finally processing the submittal. Upon receipt of the approved anchor bolt drawing from the architect, note in the same box the date when the drawing was finally distributed to the Concrete Subcontractor.

6. As you receive information requests or distribution requirements for submittals during the overall shop drawing process, include those names in the respective distribution sections.

   *a.* For example, if you receive an early request or correspondence from the Concrete Subcontractor requesting anchor bolt locations for light poles that are being provided by the Electrical Subcontractor, immediately turn to the Electrical Subcontractor's Submittal Log Form. Insert the name of the Concrete Subcontractor at "______ COPY" at the top of a distribution column and place a check next to the submittal item for the light poles. This will provide an automatic important reminder to send the submittal to the Concrete Subcontractor when it finally goes through the system.

7. Insert the dates by which the respective submittals are required. Get the information from your construction schedule and/or your purchase/award schedule. Use "early finish dates" if your scheduling system identifies such information.
8. Have the *Submittal Log* for the respective Vendor in front of you *while* you are writing the transmittal either to the design professional or back to the Vendor. After completing the transmittal, *immediately* transfer all relevant information to the Submittal Log. If you don't do it at that moment, it will never get done.

### 4.9.4 Sample Submittal Log Form—Description of Terms

The procedure for the use of the Submittal Log Form is described in Section 4.9.3. Included here are the definitions of the terms included on the actual forms provided in the following sections.

1. *COPIES.* The number of copies required to be submitted for approval that will allow *complete* subsequent distribution. Insert the actual number received.
2. *DESCRIPTION.* Title of the drawing being submitted.
3. *COMPANY DRAWING NO.* If required by the specification or by project management, submissions may be stamped to correlate with the applicable specification section or bid package, or they may simply be identified in their chronological order of submission. Both approaches can also be combined in a single numbering system as, for example:

   - Section 07210
   - 07210-2 Second submission within that section

   If no particular numbering system is used, leave the box blank.
4. *SUB DRAWING NO.* Insert the actual shop drawing number indicated in its title box whenever there is one that positively identifies the document.
5. *REQ'D BY.* Insert the date by which submission of the construction schedule is required. Use the "early submission date" if your schedule or purchase/award systems will provide such information.
6. *APPR REQ'D BY.* Insert the date by which the architect approval is required. As above, use the "early submission date" if your schedule or purchase/award systems will provide such information.

7. *FIRST SUBMISSION.* Insert the date when the respective submission is initially transmitted from the Company to the Architect for approval. The date must correspond with the date on the transmittal form used.
8. *RETURNED FROM ARCHITECT.* Insert the date when the respective submission is *received* back from the Architect after review or action.
9. *SECOND SUBMISSION.* Insert the date of any resubmission of an item requiring correction and resubmittal, either for final approval or for record. Refer back to *FIRST SUBMISSION.*
10. *ACTION/ACTION CODE.* Identify the action taken by the Architect on the respective submittal; for example, A = approved, AN = approved as noted, and so on. Add actions and codes as the specific project requires; for example, R = rejected, NM = note markings.
11. *FILE COPY.* Insert the date that a copy of the respective submission is filed into your own General File (refer to Sections 2.4 and 4.3). Every submission with any action or inaction is to be kept in the General Project File.
12. *JOB COPY.* Insert the date when a copy of the respective submission is sent to the project Superintendent for use in construction of the work. Distribute to the field *only* submissions actually to be used for construction.
13. ______ *COPY.* Insert the date when the copies of the respective submittals are transmitted to other Subcontractors and Suppliers requiring the submittal for coordination of their own work. Note the name of the Subcontractor or Supplier at the ______ COPY heading.

### 4.9.5 Sample Submittal Log Form—Completed Example *(page 4.37)*

A completed example of the Sample Submittal Log Form follows through on the procedures described in Sections 4.9.3 and 4.9.4. Section 4.9.6 provides a blank form to be photocopied for actual use.

### 4.9.6 Sample Submittal Log Form—Blank Form *(page 4.38)*

## 4.10 Submittal Requirements and Procedures

### 4.10.1 Action Responsibility

The General Contractor's interest in reviewing submittals for approval is first to confirm compliance of its Vendors and Sub-Vendors with its own agreements. Furthermore, the General Contractor and the Architect share the responsibility for shop drawing review from the project liability perspective.

The General Contractor receives, reviews, and "approves" each submittal and forwards it to the Architect and the Engineers for their approval to incorporate the item(s) into the work.

The individual responsibilities of each party to the process must be made clear in order to:

# 4.9.5
# Sample Submittal Log (Completed Example)

## SUBMITTAL SUMMARY RECORD

Project: PLAINVILLE LIBRARY
Proj. #: 9415
Owner #: N-244

Sub: QUALITY STEEL CORP.
414 LONG HILL RD.
GREENVILLE, CT 06555

Phone: (203) 555-6950
Fax: (203) 555-9950
Contact(s): JOHN IRONHEAD

Sect(s): 05100
05200
05300

**ACTION CODE:**

| | | | | | |
|---|---|---|---|---|---|
| **A** | Approved | **ANR** | Approved as Noted; Rev & Resub. | **NX** | No Exceptions Taken |
| **AN** | Approved as Noted | **NA** | Not Approved | **NM** | Note Markings |
| **FYU** | For Your Use | **R** | Rejected | **NMR** | Note Markings; Rev & Resub. |

| DOCUMENT | | DOCUMENT # | | 1st SUBMISSION | | | 2nd SUBMISSION | | | DISTRIBUTION | | | |
|---|---|---|---|---|---|---|---|---|---|---|---|---|---|
| Quant | Description | Company | Sub | Sent | Ret. | Action | Sent | Ret. | Action | File | Site | (CONC.) | ( ) |
| 9 | A.B. #1 | 05100-1 | AB-1 | 1/4/94 | 1/24/94 | NMR | 2/15/94 | | | | | | |
| 9 | A.B. #2 | 05100-2 | AB-2 | 1/4/94 | | NM | | | | 1/24/94 | 1/24/94 | 1/24/94 | |
| 9 | ERECTION DWGS | 05100-3 | E-1 | 1/11/94 | | NM | | | | | | | |
| 9 | DETAILS | 05100-4 | S-1 | 1/11/94 | | NX | | | | | | | |
| | | 05100-5 | S-2 | | | NX | | | | | | | |
| | | 05100-6 | S-3 | | | NM | | | | | | | |
| | | 05100-7 | S-4 | | | NMR | 2/15/94 | | | | | | |
| | | 05100-8 | S-5 | | | NM | | | | 1/24/94 | 1/24/94 | 1/24/94 | |

## 4.9.6
## Sample Submittal Log

### SUBMITTAL SUMMARY RECORD

Project: ______________ Sub: ______________ Phone: ______________ Sect(s): ______________

Proj. #: ______________ ______________ Fax: ______________ ______________

Owner #: ______________ ______________ Contact(s): ______________ ______________

**ACTION CODE:**

| | | | | | |
|---|---|---|---|---|---|
| **A** | Approved | **ANR** | Approved as Noted; Rev & Resub. | **NX** | No Exceptions Taken |
| **AN** | Approved as Noted | **NA** | Not Approved | **NM** | Note Markings |
| **FYU** | For Your Use | **R** | Rejected | **NMR** | Note Markings; Rev & Resub. |

| DOCUMENT | | DOCUMENT # | | 1st SUBMISSION | | | 2nd SUBMISSION | | | DISTRIBUTION | | | |
|---|---|---|---|---|---|---|---|---|---|---|---|---|---|
| Quant | Description | Company | Sub | Sent | Ret. | Action | Sent | Ret. | Action | File | Site | ( ) | ( ) |
| | | | | | | | | | | | | | |
| | | | | | | | | | | | | | |
| | | | | | | | | | | | | | |
| | | | | | | | | | | | | | |
| | | | | | | | | | | | | | |
| | | | | | | | | | | | | | |
| | | | | | | | | | | | | | |
| | | | | | | | | | | | | | |
| | | | | | | | | | | | | | |
| | | | | | | | | | | | | | |
| | | | | | | | | | | | | | |
| | | | | | | | | | | | | | |
| | | | | | | | | | | | | | |
| | | | | | | | | | | | | | |
| | | | | | | | | | | | | | |
| | | | | | | | | | | | | | |

1. Expedite the process
2. Enforce compliance with all subcontract or purchase order requirements
3. Keep relationships between Vendors and Sub-Vendors clear
4. Keep all responsibilities intact
5. Ensure that all documents are complete in all respects
6. Force prompt, correct action on the part of all parties
7. Distribute (coordinate) all information

Section 3.6 reviewed important approval responsibilities of the design professionals, and Section 4.9 developed the file system for shop drawing and submittal management. This section will provide the system to:

1. Expedite the submissions from the Subcontractors and Suppliers
2. Review each submittal for compliance with the subcontract/purchase order agreement and with the contract documents
3. Keep all Subcontractor and Supplier responsibilities with respect to the material, installation, and submittal requirements themselves where they belong

## 4.10.2 Submittal Responsibility

The shop drawing preparation and submittal procedure is normally detailed in the contract documents. Specific requirements should include size of the documents, number of copies, information required, and even some distribution requirements.

The Pass-Through Clause (see Section 3.3.9) and the opening remarks of every technical specification section will tie the Subcontractor or Supplier to the general and supplementary general conditions, which may include the overall requirements. Beyond that, the individual specification sections themselves may have specific lists of items to be submitted.

The first actual notification of specific submittal requirements should be in the subcontract or purchase order itself. The specific *date* of complete compliance should be specified therein. It is, however, in project management's best interest to continually offer assistance to a reasonable extent. Such assistance, however, should be informational only.

If any material or equipment is being submitted as "equal" or as a "substitution," do not take these categories and the contractual meaning of these words for granted. There may actually be worlds of differences in each submittal process. There may even be differences in the Owner's procedure as well. Know how the contract specifically defines these terms, along with the specific procedure for their treatment. Refer to Section 3.7, *Equals and Substitutions,* for more in this regard.

## 4.10.3 Sample Letter to Subcontractors Regarding Submittal Requirements *(page 4.40)*

The Sample Letter to Subcontractors Regarding Submittal Requirements that follows will first help you to increase your own familiarity with the technical requirements and highlight them for your Subcontractors. It will make it easier for your Subcontractors to be aware of the requirements, and of the fact that the Company

## 4.10.3
## Sample Letter to Subcontractors Regarding Submittal Requirements

**Letterhead**

Date: ____________________

To : ______________________________
______________________________
______________________________

ATTN: ______________________________

RE: Project:____________________
Project #:____________________

SUBJ: Submittal Requirements
Section(s)____________________

Mr.(Ms.)________________________:

The contract requires your submission of the following items. Note that this list is for your convenience only; the omission here of any item does not relieve you of the requirement.

Please submit:

___ Certificate of Insurance
___ Performance Bond
___ Labor & Material Payment Bond
___ Certified Payroll Reports
___ Payroll Ledgers
___ Shop Drawings:____Copies
___ Erection Drawings
___ Product Specifications
___ Employer EEO Reports
___ MSDS Sheets
___ Samples
___ Installation Instructions
___ Delivery Time After Approval
___ Erection/Installation Time
___ Tests/Inspection Reports
___ Guaranties/Warranties
___ Full Lien Waiver Forms
___ As-Built Plans & Specs
___ ____________________
___ ____________________
___ ____________________
___ ____________________

Please be aware of the proper form, content, and time requirements for each submission, and comply in every respect. Contact me immediately if you have any questions.

Thank you for your cooperation.

Very truly yours,

COMPANY

______________________________
Project Manager

cc: File:__________, CF

takes them very seriously. The letter will later serve as a convenient checklist for follow-up.

The letter is to be filled out and sent to each Subcontractor and Supplier on or about the time that the respective subcontract or purchase order is sent. It will thereby:

1. Call attention to all required submittals
2. Serve as first notice that the party is absolutely expected to comply with all requirements
3. Serve as additional notice (besides the subcontract itself) that the party is responsible for all information contained in the plans and specifications
4. Advise of where the party can get all relevant information (plans and specs) in order to avoid any possible claim that they don't have them available
5. Serve as notice that all approved submittals of contiguous work required by the Subcontractor and Supplier for coordination of their submittals are available for inspection at the jobsite (your failure to have directly sent that Subcontractor its own copy is therefore no excuse)
6. Require that your job number be included to help with your own file procedures

In order to use this letter:

1. Review the Owner's specification requirements for all submittals as they are included in the respective subcontract or purchase order. Indicate them in the letter.
2. Review the subcontract or purchase order and its standard and special conditions. Indicate these special requirements in the area provided in the letter.
3. Send the letter, making sure to send copies to the Site Superintendent and other relevant project personnel.

### 4.10.4 Submittal Review Checklist
*(page 4.42)*

It is project management's responsibility to review the shop drawings, product data, and samples to confirm their compliance with the criteria set forth in Section 4.10.3. The actual process performed by those individuals responsible should become a matter of detailed routine.

Using a checklist each time will help you to:

- Avoid overlooking important considerations
- Make certain that all approval submissions meet all requirements
- Force prompt, correct action on the part of the design professionals
- Coordinate each in a timely manner with all those who need the information
- Follow up on resubmissions and noncompliances

Beyond this, it is a helpful practice during your reviews to use markers of different colors than the design professionals (typically red). You might even suggest to the designers that they use different colors among themselves. This will make it very easy later on to confirm the origination of any remark.

# 4.10.4
# Submittal Review Checklist

**SUBMISSION REQUIREMENTS**

All approval submissions contain:

1. Project Title & Job Number ____
2. Contract Identification ____
3. Date of Submission (or Revision) ____
4. Dates of Previous Submissions ____
5. Names of contractor, supplier and/or mfgr. ____
6. Identification of all products with specification section numbers ____
7. Field dimensions - clearly identified as such ____
8. Relation to adjacent and/or critical features of the work ____
9. References to applicable standard specs ____
10. *Clear* identifications of deviations from the Contract Documents ____
11. All other pertinent information as may be required by the specifications or the Company, such as:
    - Model Numbers ____
    - Performance Characteristics ____
    - Dimensions & Clearances ____
    - Wiring or Piping Diagrams ____
12. Manufacturer's standard drawings include:
    - Modifications to delete information not applicable to this project ____
    - Supplemental information specifically applicable to this project ____
13. Check the specifications for add'l requirements ____

**SUBMITTAL REVIEW PROCEDURE**

1. Ensure that subcontractors and suppliers submit materials promptly ____
2. Determine and verify:
    - That the sub has incorporated and will guarantee all field dimensions ____
    - All field conditions and construction criteria have been accommodated ____
    - That the product either complies with the specification requirements in every respect, or that every deviation has been properly identified, and includes its respective complete explanation/justification ____
3. Coordinate each submittal with both field and contract document requirements ____
4. Research and confirm all "justifications" for any deviation from the contract documents. Do this *before* submitting the documents for approval ____
5. Determine if a credit or addition to the contract is in order, based upon any changes in the submission ____
6. Determine if any backcharges to any other subcontractor or supplier are in order as a result of changes required by this item ____
7. Determine that the submission is timely, and that the material conforms to required deliveries ____
8. Positively identify by responsibility all "Not By Subcontractor" or "By Others" kinds of remarks. Correct as necessary *before* submission to the architect for approval. ____
9. Compare all resubmissions with the file copy of the previous submission. Confirm that all required corrections have been made. ____

**DISTRIBUTION**

1. Upon receipt of submittals bearing the stamp indicating architect action, distribute copies to:
    - Jobsite File ("For Construction" documents only) ____
    - Record Documents File ____
    - Other affected subcontractors and suppliers ____
    - The supplier or fabricator ____
    - Anyone else who may need the information in order to coordinate the work properly ____

**FOLLOW UP**

1. Monitor the time that it takes for the approval process.
    - Be sure that the architect is giving proper, timely attention ____
    - Be sure that all delays and other inappropriate actions are duly noted in the correspondence ____
2. Be certain that the design professionals:
    - Include all information required of them by way of questions in the submittals ____
    - Do not overstep their authority ____
    - Do not overstep their professional capacities ____
    - Do not add work without regard for the established change order procedure ____
    - Include only meaningful action that will allow proper completion of the submittal ____
    - Affix the *accepted* stamp and initial/sign it ____
    - Clearly indicate any requirements for resubmittal, or approvalof the submittal ____
3. Upon distribution of the submittal back to its originator:
    - Reconfirm the delivery schedule(s) ____
    - Confirm that the submission is being returned in good time for the subcontractor or supplier to meet its own requirements ____
    - Note any significant information for the next construction schedule update ____
    - Begin any actions that may be necessary to resolve problems that may have been exposed by the review process ____
    - Begin any necessary change order procedure ____
4. Be sure that the Submittal Log form is used and maintained *as each part of the process is completed*. Complete the respective log entry

### 4.10.5 Sample Form Letter to Subcontractors Regarding Shop Drawing Resubmittal Requirements *(page 4.44)*

The Sample Form Letter to Subcontractors Regarding Shop Drawing Resubmittal Requirements that follows is to be used to follow up on those submittals that for whatever reason have been returned to the Subcontractor or Supplier for correction and resubmittal for final approval. It is an effort to compel an unresponsive Subcontractor or Supplier to devote proper attention to *timely* compliance. Specifically:

1. It "reminds" the recipient to devote more attention to the project. It makes them aware of the critical nature of their prompt action, and that you're not about to relax until the process is completed.
2. It documents your efforts to coordinate the work. These kinds of records will become extremely important in the event that untimely or inappropriate action on the part of the Sub-Vendor results in slipping material deliveries. It will also provide support for backcharges or other more serious claims.
3. It notifies the party of potential backcharges that may result from inattention to the project.

To use the letter effectively, review the Submittal Log of Section 4.9 periodically. Immediately send out the letter to each Subcontractor and Supplier who are delinquent in their responses to the resubmission requirements. Sending "second-request" and "third-request" copies of the original letters in case of continuing lack of response prompts more drastic action.

### 4.10.6 Reproductions of Submissions—Sub-Vendor Responsibility

The proper quantity of submittal documents is specified in the subcontract, the purchase order, or elsewhere in the contract documents. On a public project, for example, it will be common to require sufficient numbers of submittals that will provide for final distribution to:

- The Architect (home and field?)
- The Owner's administrative and field offices
- The Company's home and field offices
- The Company's general file
- The Subcontractor's or Supplier's home and field offices

Beyond this, any engineer or consultant will need one or two copies, and all Subcontractors needing the submittal for coordination of their own work will need an office and a field copy.

It is easy to see, for example, that it may be necessary to have up to fifteen copies of the HVAC duct shop drawing. Even the simplest submissions may need upward of ten copies for only basic contract distribution.

Too many Subcontractors and Suppliers, however, take it upon themselves to submit only a few copies of each submittal—obviously because of the inconvenience and cost. If this abuse is allowed, it will interfere with the otherwise smooth flow of document distribution, and will greatly add to the cost of reproductions for the Company.

## 4.10.5
## Sample Form Letter to Subcontractors Regarding Shop Drawing Resubmittal Requirements

**Letterhead**

Date: ______________________________

To: ______________________________
______________________________
______________________________

ATTN: ______________________________

RE: Project:______________________________
Company Project #: ______________________

SUBJ: Shop Drawing Resubmission Requirements
Section(s) ______________________________

Mr. (Ms.) ______________________________:

Copies of your shop drawings *requiring immediate correction and resubmission for approval* have been returned to you on the dates listed below. Immediate resubmission is required in order to avoid additional delays and related costs resulting from untimely action.

| Drawing No. | Date Returned to Your Office | Description |
|---|---|---|
| | | |
| | | |
| | | |
| | | |
| | | |
| | | |
| | | |
| | | |

Please contact me immediately to confirm the date(s) of your resubmissions(s). Time is of the essence.

Thank you for your cooperation.

Very truly yours,

COMPANY

______________________________
Project Manager

cc: Jobsite
File: Vendor File________
CF

It is project management's responsibility to enforce each submittal requirement. The most effective way to this enforcement is to assign a cost to the service.

Because the objective is to enforce the Subcontractors' and Suppliers' compliance and *not* to go into the copy business, the charges assigned to the effort should intentionally be very high. They should raise a little anger and shock the offending party into compliance. It should become quickly evident to them that it is much less of a problem to just fulfill their responsibilities than to argue over large backcharges.

### 4.10.7 Sample Reproduction Backcharge Notice Form Letter *(page 4.46)*

The Sample Reproduction Backcharge Notice Form Letter follows through on the ideas of Section 4.10.6. Specifically, it:

1. Highlights the contractual requirements for proper, complete submissions
2. Notifies of the intention to levy backcharges and associated costs

After having sent out this letter, be prepared for an immediate and abrupt phone call. Stay calm and simply advise the party that their failure to provide the correct submission is truly an interference. They can avoid these charges by simply delivering what they're supposed to.

## 4.11 Subcontractor Delivery Requirements

### 4.11.1 General

After the approval process has been completed, focus must be shifted to actually securing the materials within acceptable time frames. This section reviews techniques to expedite the delivery of materials and enforce compliance with all Subcontractor and supplier delivery requirements.

### 4.11.2 Expediting Subcontractor/Supplier Fabrication and Delivery Schedules

Begin with the attitude that it is *always* necessary to expedite *everything.* The construction business is founded on the squeaky-wheel concept. If your wheel isn't squeaking, you can be sure that all the other wheels with which your Subcontractors are working are. If you keep *your* wheels squeaking, you'll get *your* material on time.

This section identifies considerations and activities that may be applied in various circumstances to help avoid delivery problems in the first place, or to help bring problem deliveries back on line. They are not to be considered to be in any particular order, but as a menu from which to select appropriate actions.

1. *Identify the true source.* Begin with the subcontract itself to require the disclosure of all actual sources of all materials and components. Continue with the *Subcontractor / Supplier Reference Form* of Section 4.5.

# 4.10.7
# Sample Reproduction Backcharge Notice

**Letterhead**

Date: ______________________

To : ________________________________
________________________________
________________________________

ATTN: ________________________________

RE: Project:________________________
Project #:______________________

SUBJ: Failure to Provide Proper Content of Submissions

Mr. (Ms.)___________________________:

Per your contract, it is your responsibility to be sure that the proper number of copies for all required submittals be sent on time to be processed. Failure to comply interferes with the expedient flow of documents, and results in significant additional production costs.

Be advised that all copies made by this office of any of your submissions because of insufficient quantities necessary for proper distribution will be charged to your account as follows:

8-1/2" x 11" . . . . . . . . . . . . . . . . . . $ 1.00 Each Leaf
8-12" x 14" . . . . . . . . . . . . . . . . . . $ 2.00 Each Leaf
11" x 17" . . . . . . . . . . . . . . . . . . $ 3.00 Each Leaf

Prints:
18" x 24" . . . . . . . . . . . . . . . . . . $ 5.00 Each Leaf
24" x 36" . . . . . . . . . . . . . . . . . . $ 8.00 Each Leaf
Larger . . . . . . . . . . . . . . . . . . . $ 10.00 Each Leaf

Before final payment is released, your account will be adjusted to reflect the number of copies charged.

Very truly yours,

COMPANY

________________________________
Project Manager

cc: Jobsite
Vendor File:________
Backcharge File
CF

2. *Establish contact with the direct source as early as possible.* Know early whom to call to directly verify all information handed to you by your first-tier Subcontractor. Develop the relationship as soon as possible.
   - *a.* Explain that if Vendors want to be sure that they get paid properly, it is in their best interest to communicate directly with you:
     - (1) *You are* the one with the payment bond.
     - (2) *You are* the one who will be processing payments, with the power to determine joint-payment structures.
   - *b.* In most cases your immediate result should be that the reliability of your delivery information will improve dramatically. You'll also be able to use this direct information as a gauge to help determine the reliability of the delivery information for other items that you'll be getting from the same first-tier Subcontractor.
3. *Confirm the relationship history of the Sub-Supplier with the first-tier Subcontractor.* You may, for example, discover that there really is no relationship to speak of, and the Supplier is really very happy to be talking with you.
4. *Understand the specific payment terms between the Subcontractor and its Supplier.* Too often they are dramatically different than those between you and your Subcontractor. Even though it is not technically your problem, these cases can have a tendency to introduce additional conflicting forces between you and your Subcontractor.
   - *a.* One consideration may be to offer Sub-Suppliers a copayment agreement in return for allowing them a pay-when-paid. This might go a long way in relieving pressure during normal production cycles.
5. *Consider visiting the fabrication location.*
   - *a.* When you're told "the parts are on the shop floor," consider responding with "great; I'll be there this afternoon—you can show them to me." You'll thereby either truly satisfy yourself that the information is real, or you will have exposed an all too common "check-is-in-the-mail" syndrome of this industry. At the very least, you'll get your Vendors to consider their answers to you more carefully.
   - *b.* Even when all information is absolutely real, just the act of visiting a remote location to see your items communicates your genuine concern for the efforts and hard work on the part of that Vendor to get you your things on time. Many times it may need nothing more than this appreciation to give just the added impetus that keeps *your* material fabrication moving along. Without any doubt, shop visits have been *the* most powerful and consistently successful of all expediting efforts.
6. *Point out that if the item is delivered by the end of the month, it will go on that month's requisition to the Owner.* In other words, if the item comes a week later, everyone is going to have to wait an additional 30 days to even submit the invoice.
7. *Identify whether the shipment is arranged direct or by common carrier.* Many shipments either are waiting for a full truck or will be consolidated by common carrier. It is often a reasonable upcharge to get the materials on their own direct truck.
8. *Consider accelerated payments.* If, for example, the Supplier is on a 30-day or even a pay-when-paid payment term, consider offering a c.o.d. or other advanced form of payment. Don't necessarily give it away, however. Even

though it will speed up delivery, it may also be justification for further price discount.

9. *Consider advising the Vendor that you're considering changing the item.* If the delay is the fault of the Vendor, consider doing what you can to develop the idea that they might lose the order altogether. Do this with extreme caution. Make it believable, and *actually be ready to do it,* or you might actually be digging a deeper hole if the Vendor drops *you.*

Throughout these efforts, those that become necessary as a result of the primary Subcontractor's failure will be chargeable to that Subcontractor under the acceleration and backcharge provisions of the subcontract agreement. Refer to Section 4.16, *Backcharges,* for more in this regard.

### 4.11.3 Sample Subcontractor/Supplier Delivery Requirements Form Letter

The Sample Subcontractor/Supplier Delivery Requirements Form Letter is designed to provide a record of proper notification of required deliveries. It should become a matter of habit to return the individual submission of the shop drawing to the Subcontractor after approval action to indicate expected delivery and request positive confirmation (see Section 4.6, *Transmittal Form Letter Procedure and Use*). If this had been done, the Delivery Requirements Letter will become at least the *second* such notification.

If acknowledgment is not received by the date indicated in the letter, stamp a copy of the letter with "Second Request" and indicate on the second letter that failure to respond is interfering with project coordination.

## 4.12 Request for Information

### 4.12.1 General

The Request for Information (RFI) will actually become either a request for information or a request for confirmation of a clarification. It is a procedure to be used in cases where it is necessary to:

1. Confirm your interpretation of a detail or other understanding as to the way a component of the work should proceed
2. Secure the written direction or other clarification from the appropriate party that is necessary in order to allow the work to proceed

The Request for Information becomes a crucial element of the complete project record. Learn to express your thoughts on it, and your project documentation will improve dramatically.

### 4.12.2 Use of the RFI Form Letter

1. The RFI is to be used in all cases for information clarification involving the design professionals, Subcontractors, Suppliers, and Owner representatives. Use the chronological order of RFI initiation as the RFI #.
2. Summarize the issue in section 1 of the RFI. Be specific; refer to plan details, specification section numbers, and so on. If it is a confirmation of a conversation or other direction, specifically state it. Name names; include dates.

## 4.11.3 Sample Subcontractor/Supplier Delivery Requirements Letter

**Letterhead**

Date: ______________________

To : ________________________________
________________________________
________________________________

ATTN: ________________________________

RE: Project:______________________
Project #:______________________

SUBJ: Delivery Requirements
Section(s):______________________

Mr. (Ms.)__________________________:

Copies of "Approved" or "Approved as Noted" Shop Drawings were returned to you on the dates listed below.

Per the current construction schedule, delivery of each item is required on the respective dates indicated in order to avoid delays and related costs resulting from nondelivery.

| Drawing No. | Description | Date Returned to Your Office | Required Delivery Date of Material |
|---|---|---|---|
| | | | |
| | | | |
| | | | |
| | | | |
| | | | |
| | | | |
| | | | |
| | | | |
| | | | |

Please confirm your ability to meet these delivery dates by signing this letter where indicated below and returning it to my attention. If any item will not be delivered by the required date, submit your complete explanation. Your complete response is required by __________________, 19______.

Very truly yours,

COMPANY

______________________________________
Project Manager

Delivery Dates Confirmed:

By:______________________Date:____________

3. Do the best you can to actually confirm the answer to the question first. This, of course, will greatly speed up your ability to proceed with the work. The form will then be used as confirmation.
4. If you are unable to confirm the answer *now,* send the form with the written statement of the issue to the responsible party with section 3 checked. Use the fax, hand delivery, or Express Mail if necessary. Do whatever it takes to get the form into the hands of the other party immediately. Stress the urgency of a quick but complete response.

### 4.12.3 Sample Request for Information (RFI) Form Letter—Completed Example *(page 4.51)*

### 4.12.4 Sample Request for Information (RFI) Form Letter—Blank Form *(page 4.52)*

### 4.12.5 Use of the RFI Tracking Log *(page 4.53)*

The RFI Tracking Log can be kept in its own section of the binder being used for the *Submittal Log* of Section 4.9, adding additional log forms as necessary.

1. Log the information *as it occurs.*
2. Indicate the chronological order of the RFI as the RFI #.
3. Complete all "initiation" information as of the RFI preparation:

| | |
|---|---|
| TO | The recipient. |
| DATE | Date of the RFI. |
| CONFIRM | Check if the RFI is confirming a clarification or direction. |
| ACTION REQUEST | Check if the RFI is requesting direction. |

4. Periodically (often) review the log to identify outstanding information or outstanding written acknowledgments of your confirmations.
5. Routinely send "Second Request" copies of the original RFI.
6. Complete the ACTION REC'D information as it is received and keep the log current at all times.

## 4.13 Project Meetings

### 4.13.1 General

Project meetings are generally divided between regular periodic job meetings and special meetings held to address specific unique problems or circumstances. In either case they should be treated the same in terms of their arrangement, participants, rules of conduct, and documentation.

The discussions of this section will apply to all meetings, in particular regular job meetings. Additional considerations applying to special meetings are developed in Section 4.13.14.

## 4.12.3
## Sample Request for Information (RFI)
## (Completed Example)

**Letterhead**

Date: JULY 22, 1994 **REQUEST FOR INFORMATION: RFI #:** 16

To: STATE ARCHITECTURAL UNIT
165 BURECRATIC BLVD.
CITY, STATE 00000

Project: BURLINGTON POLICE TWR
Project #: 9550

____ Hand Delivered X Mailed

ATTN: JOHN ARCH

X Faxed: 666-9999

Subject: Section (s): 03200 CONCRETE FOUNDATION LAYOUT
Specification/Plan References: S-1, A-4

1. Problem/Information Requested:
DWG S-1 DET. 6 REQUIRES THE FROST WALL AT COL. LINE 3 TO BE 14" THICK.

5/A-4 REQUIRES THE SAME WALL TO BE 12" THICK (IN ORDER TO WORK PROPERLY WITH THE BRICK SHELF)

Information Requested By: MARK LEONARDO

2. Response Confirmation:
YOU CONFIRMED IN OUR 9:40 AM TELECON TODAY THAT DETAIL 5/A-4 WILL BE FOLLOWED
(6/S-1 WILL BE CORRECTED ON THE AS-BUILT DRAWINGS)

Approved By:____________ Date:________

3. Clarification/Action; please respond by or before ____________, 19____ in order to minimize delay or interference with the ability to proceed with the work:

By:____________ Date:________

cc: Superintendent
OWNER ON-SITE REP
RFI #16
File: CONC, CF

## 4.12.4
## Sample Request for Information (RFI)

| Letterhead |
|---|

Date: ______________________ **REQUEST FOR INFORMATION: RFI #:___**

To: ______________________ Project: ______________________
______________________ Project #: ______________________
______________________

_____ Hand Delivered _____ Mailed
ATTN: ______________________ _____ Faxed: ______________________

Subject: Section (s): ______________________
Specification/Plan References: ______________________
______________________

1. Problem/Information Requested: ______________________

Information Requested By: ______________________

2. Response Confirmation: ______________________

Approved By: ______________________ Date: __________

3. Clarification/Action; please respond by or before ______________________, 19_____ in order to minimize delay or interference with the ability to proceed with the work:

By: ______________________ Date: __________

cc: Superintendent
______________________
______________________
File: ______________________, CF

## 4.12.5

## Sample RFI Tracking Log

# RFI TRACKING LOG

**Project:** ______________________________ **No:** ____________

| RFI # | SUBJECT | INITIATION | | | | ACTION RECEIVED | | REMARKS |
|---|---|---|---|---|---|---|---|---|
| | | TO | DATE | CONFIRM | ACTION REQUEST | Complete ? | DATE | |
| | | | | | | | | |

### 4.13.2 Meeting Purpose

Job and special meetings are there to *solve problems*—not just to rehash the same items that you've discussed last week. These meetings are critically important to the quick and complete resolution of every item affecting the project—if they are *managed* and not avoided. They are not simply there to record history, but to force action, pinpoint accountability, and support your actions.

### 4.13.3 Day and Time

Mondays and Fridays will be sparsely attended. Even if you can manage participation, attention spans will be divided, and you'll never be able to maintain proper consistency from meeting to meeting without constant bird-dogging of all required participants.

Tuesdays or Thursdays are best. Whenever the rest of the world schedules anything, it is sure to be on Wednesday, creating the inevitable conflict. Wednesdays do, however, remain much better than Mondays or Fridays.

Insist on morning meetings, starting between 10:00 and 11:00 A.M. This will:

- Allow a few precious minutes for last minute preparations
- Give people a chance to get to the meeting without having to fight morning or noon-hour traffic
- Catch attendees before the rest of the day is allowed to interfere with their schedules, thereby improving meeting attendance
- Leave the rest of the day to act on critical issues before they're allowed to cool
- Improve your chances of catching someone else "in" if you need to phone or visit them today to resolve something
- Leave time to force others to change *their* plans for the rest of the day in order to resolve a current problem
- Keep everyone working toward getting through the agenda and finishing the meeting

The closer you get to lunch, the quicker things seem to get resolved. In marked contrast, meetings held after lunch are conducted in slow motion. They're sure to be sparsely attended, move at half-speed, and close with most of the original agenda still intact. You can't have a full head and a full stomach at the same time—one of them has to be empty.

### 4.13.4 Location

*Always* meet at the site. Don't waste time arguing over whose project it is, or get caught in the one-upmanship game of showing would-be authority by fighting over meeting location. Heads of bureaucratic government agencies, for example, have a habit of trying to arrange for all regular and special meetings at their own offices or at some other location convenient for them.

The jobsite is where the issues live. Get out of the field office, walk to the problem, point at it. Misunderstandings will clear up in seconds, and you'll move closer to resolution.

### 4.13.5 Participants

The absolute minimum will be the Owner's representative(s), the Architect, construction field personnel (Superintendent, Project Manager), and construction administration (Project Manager, Project Engineer). It unfortunately is getting to be too common on small and mid-sized projects for the mechanical, electrical, structural, or other special engineers not to attend, usually because the Architect retains these consultants on an hourly basis during the construction phase of the project. That is *their* problem, not yours. If you need or want the engineer present for any valid reason (including expediency), get him or her there.

The arrangement may otherwise work if you have no engineering problem (?!). If you do, however, their lack of attendance is your guarantee that a *minimum* of a week will be added to any related design resolution. If your agenda has design-related problems, insist on these individuals' attendance. If they then do not attend, you will have every right to call foul and highlight the potential schedule impact resulting from the extra delay in resolution.

### 4.13.6 Subcontractor Participation

Job meeting attendance is *mandatory* for all Subcontractors who:

- Are about to work
- Are working
- Just "finished" working
- Have any potential for being involved in any agenda item

Too many times, Subcontractors are allowed to avoid meetings because of the realization that they will have to sit there for an hour (maybe two) until their item comes up. Don't allow it. If a Subcontractor affected by any discussion is not present:

1. Issue resolution is delayed.
2. Extreme amounts of effort will be necessary to coordinate subsequent discussions between multiple parties.
3. *You* will remain responsible for the timely coordination of all information to those who are not there.
4. Even if you conduct proper coordination and distribution of relevant information, this adds significant time and effort.

In contrast, the time inconvenience of the individual Subcontractors is a small price to pay when compared with the fact that:

1. The issues were coordinated and resolved in minutes with all affected people present
2. There is no time lapse in your coordination, or with your ability to immediately mobilize any subsequent action

Refer to Sections 4.13.7 and 4.13.8 for help in inducing such attendance.

### 4.13.7 Sample Letter to Subcontractors Regarding Mandatory Job Meeting Attendance *(page 4.56)*

The Sample Letter to Subcontractors Regarding Mandatory Job Meeting Attendance that follows is designed to induce Subcontractor participation in your meetings. Specifically, it notifies all project Subcontractors that:

## 4.13.7
## Sample Letter to Subcontractors Regarding Mandatory Job Meeting Attendance

**Letterhead**

(Date)

To: (List all Project Subcontractors)

RE: (Project)
(Company Project #)

SUBJ: Mandatory Job Meeting Attendance

Mr. (Ms.) ( ):

Your contract requires your participation in the regular job meetings. These meetings will begin on (Date) and will be held on alternating (insert day of the week). This schedule may be adjusted from time to time; it is your responsibility to be aware of the current job meeting schedule.

Each subcontractor is required to attend every job meeting prior to, during, and immediately after the work of the subcontract is being performed. This includes the submittal preparation/submission stage.

Please note that your attendance is *mandatory*. Your failure to attend will result in the need for excessive efforts by others to coordinate their work with yours. You will be held responsible for all information contained in the meetings, including timetables, commitments, and determinations of responsibility as set forth.

Thank you for your cooperation.

Very truly yours,

COMPANY

Project Manager

cc: Jobsite
File: Meetings
CF

1. It is mandatory that anyone performing or about to perform *any* work on the site is absolutely required to participate in all job meetings during that period.
2. Meetings will be held on the dates scheduled and will start *on time*. Attention is expected to be given to these requirements.
3. It is each Subcontractor's responsibility to be aware of all information as it relates to its work, and to make all efforts necessary to ensure proper coordination.
4. Each Subcontractor is absolutely responsible for all information contained in the job meeting minutes. This includes completeness, accuracy of description, noted commitments, and timetables.

### 4.13.8 Sample Letter to Subcontractors Regarding Lack of Job Meeting Attendance *(page 4.58)*

The Sample Letter to Subcontractors Regarding Lack of Job Meeting Attendance is designed to deal with a particular Subcontractor who does not properly attend your meetings. Specifically, it confirms your conversation with that Subcontractor who, despite your coordination efforts, has failed to attend your meetings.

1. It emphasizes that the lack of attention is creating unnecessary interferences and inconveniences.
2. It states that interferences, delays, and additional costs resulting from the lack of attention will be entirely that Subcontractor's responsibility.
3. It reiterates that it continues to be the *Subcontractor's* responsibility to be aware of all project requirements as determined in those meetings, and to comply with them in every respect in a timely manner.
4. It "reminds" the Subcontractor of the next job meeting.
5. By copy of the letter it makes certain that the field representative's boss is aware of the absence and your feelings toward it. If that doesn't get some reaction, then you probably have other problems with that company.

### 4.13.9 Meeting Action Rules

1. Schedule meetings as *you* need them to be.
   *a.* During the job start-up, shop drawings are flying, construction and bid package coordination efforts are being compressed into small time frames, and resulting questions are multiplying.
   *b.* Insist on weekly meetings during these aggressive periods. Too often those responsible for acting on a job meeting item will not look at the item until the day before (or the morning of) the next job meeting. Deadlines seem to be the earliest date that you can hope for any action. Get these deadlines (the next meeting) as close together as possible.
   *c.* As the project settles into a pace, it's up to you if you think it will be OK to "relax" into biweekly meetings. If, however, you get any hint that too much time is spent between issue identification and resolution, immediately get back to the weekly schedule.
2. *Always* start on time—regardless of who is late and of how many times that person has been late.

## 4.13.8
## Sample Letter to Subcontractors Regarding Lack of Job Meeting Attendance

| Letterhead |
|---|

(Date)

To: (Subcontractor failing to)
attend a specific meeting)

RE: (Project)
(Company Project #)

SUBJ: Lack of Job Meeting Attendance

Mr. (Ms.) ( ):

Per our conversation this date, your failure to attend today's job meeting as required is interfering with job coordination and completion. As you know, it continues to be your responsibility to be aware of all project requirements and to accommodate them completely and in a timely manner. Please be advised that you will be held responsible for all interferences, delays, and added costs resulting from this lack of attention.

The next job meeting will be held on (insert day and date) promptly at (insert time).

Very truly yours,

COMPANY

Project Manager

cc: Jobsite
File: Vendor File:__________
Meetings
CF

  *a.* One or two times having to sneak into an ongoing meeting will usually solve the problem. If it does not, chronic offenders should be confronted at the meeting. Let those who do get there on time know that you appreciate their efforts, and that *none* of you appreciate the lack of consideration being demonstrated.
  *b.* If for some reason you cannot begin (the Owner's representative is driving in from another city, and the design professional refuses to start...), consider stating at the actual start of the meeting that you now need an adjustment of the agenda to hit the important topics because *you* must leave on time. *Get and keep control over the meeting. Let everyone know that your time is valuable, and that they need to learn to respect that. You've managed to keep your commitments, you have a right to expect others to do the same.*

3. Enforce mandatory attendance.

  *a.* Do not tolerate absence or neglect. When an expected attendee is missing, it disrupts the agenda and loses time. When the agenda item comes up, phone the person right in the middle of the meeting. If a speaker-phone is available, put him or her on it. Preface your conversation with "we expected you here, but since you're not, we've got you on the speaker, so..." and move right into the issue. A mild reprimand for their lack of consideration (a strong one for repeated offenders) is definitely appropriate. Be matter-of-fact and businesslike. These people will be caught at least a little off-guard, and hopefully be embarrassed enough to be sure they won't put themselves through it again.

4. End each item with a *resolution.* If the issue itself is not finalized, end with a determination of a specific action to be made by a particular individual (by name) by a certain date. Nail it down.
5. Ongoing, comprehensive records must be kept in a way that keeps:

  *a.* The project record clear and correct
  *b.* Everything on the front burner
  *c.* Everyone accountable to and for his or her actions (and inactions)

Refer to Section 4.13.10, *Meeting Minutes,* for related action rules.

## 13.10 Meeting Minutes

All meetings must at least be recorded, but that in itself is not nearly enough. If the meeting is conducted properly:

- Each item is given its relevant place on the agenda (see Section 4.13.13).
- The *timing* of each item can be controlled.
- Each issue has either been completely resolved or ongoing items have had definite steps determined for resolution, along with naming those responsible and confirming action timetables.

The minutes themselves will:

1. Organize the agenda
2. Establish the method of identification and correlation to the rest of the project record
3. Keep everything *visible*—and organized so that unresolved items automatically get more irritating the longer they go unresolved

4. Nail down accountability, that is, keep people directly identified and *personally responsible* for specific actions
5. Display the cause and effect of timely and untimely actions
6. Fulfill important notification responsibilities under the contract (see Section 3.3.16)

The following guidelines should be followed in recording the minutes:

1. *Use a standard form.* Whether on a word processor or kept on standard forms made for the purpose, a standard layout should:

   - Display the project identification, meeting number, date, time, location, participants, distribution, and other requirements unique to the project
   - Prompt the recording of all relevant information while reducing the risk of oversight
   - Get everyone used to the information display

   This will improve the understanding of the issue identification and correlation features of the minutes and minimize the risk of someone overlooking his or her responsibilities or otherwise claiming ignorance.

2. *Identify each meeting numerically.* If they are regular job meetings, identify each meeting in numerical sequence. If it is a special meeting, call it Meeting #1.
3. *Assign each item its own number that will never change.* The third issue raised at job meeting #4, for example, will be identified as "4.3."

   - If you are conducting weekly job meeting #8, for example, and you find yourself still considering item 4.3 under "Old Business," you automatically know that the issue is four weeks old.
   - If you are a supervisor who does not attend the regular job meetings, simply reviewing any meeting minutes and comparing the item numbers with the meeting number will give you an instant (and sometimes painfully clear) indication of the way things seem to be going.

4. *Use a title for each item, and keep titles consistent each time the item is mentioned.* This will clarify the subject, speed research, and facilitate correlation with other topics.
5. *Include all appropriate references in the job title.* If it is the subject of a change order, bulletin number, etc., keep these identification numbers in the description.
6. *Be concise but complete.* Use outline format whenever possible.
7. *Require definite action.* Never leave any issue without the specific step-by-step program identified that will resolve it. Assign people and times to each step.
8. *Name names.* Do not say "the Owner will respond by...." Instead, say "Mr. Dunn stated...." Let everyone see their names in lights—it will be harder to make excuses.
9. *Insist on the precise accuracy of all statements as they are recorded.* If you are not keeping the minutes yourself, keep precise notes. If the minutes either represent an issue inaccurately or omit relevant discussion, highlight the correction at the next meeting. If it is not a regular meeting, immediately distribute your written correction.

10. *Include a "verification requirement" of all information contained in the minutes.* The minutes are an important job record that will be used to substantiate every cause-and-effect issue. If its accuracy is questioned, its usefulness will be compromised. At worst it may pervert the record in a manner that will hurt you and the Company.
11. Include as part of the standard form on every meeting record a statement to the effect that anyone noting any error or omission in the document is to notify the writer by or before the next meeting, or by a particular date, and that failure to do so constitutes acceptance of all information contained as it is represented. This will end many later arguments regarding the legitimacy of particular remarks.

### 4.13.11 Sample Job Meeting Minutes Form—Pages 1 and 2 *(pages 4.62 and 4.63)*

The Job Meeting Minutes Form that follows is arranged to accommodate the requirements that must be provided for at each meeting. In addition, the form accommodates the objectives of this section by:

1. Providing appropriate areas to prompt the inclusion of all relevant job meeting identification and distribution information
2. Encouraging documentation in accordance with the "outline form" recommendation
3. Providing a convenient area for highlighting important necessary action, including person responsible and date required by
4. Including the important notification to correct any errors or omissions in the record of noted discussions

### 4.13.12 Sample Meeting/Conversation Record Form *(page 4.64)*

The Meeting/Conversation Record Form that follows is to be used where:

1. A verbal exchange has been more elaborate than should be included as only a note in the *Telephone Log* of Section 4.4, but not as elaborate as a formal meeting.
2. It is necessary to confirm instructions, notification to proceed with work in a certain way, or other distribution of information.
3. It is important to confirm the appropriate record of discussion to the participants and to others, to properly document the actual project record with regard to the subject items, and to verify the legitimacy of your record of the exchange.

Accordingly, the main difference between using this form and simply elaborating on your own notes in the Telephone Log is the distribution and confirmation of the information.

### 4.13.13 Meeting Agendas

1. *Concept.* If you control the agenda, you will control the timing and content of all discussions. You will also control what will and, perhaps more importantly, what will not be said. Agendas can:

## 4.13.11
## Sample Job Meeting Minutes Form
## (Page 1 of 2)

Project:
Proj. #:

Job Meeting No.:
Date:
Location:

Page 1 of

| ATTENDING: Name | Company | DISTRIBUTION Name | Company |
|---|---|---|---|
| | | Attendees: | |

NOTICE TO ATTENDEES AND MINUTES RECIPIENTS:
If any of the following items are incomplete or incorrect in any way, please notify the writer. Failure to advise of such corrections by or before the next job meeting constitutes acceptance of all information contained herein as represented.

| SUBJECT | ACTION REQUIRED | |
|---|---|---|
| | By | Date |
| | | |

# 4.13.11
# Sample Job Meeting Minutes Form
# (Page 2 of 2)

Project: Job Meeting No.: Page of
Proj. #: Date:
Location:

| SUBJECT | ACTION REQUIRED | |
|---|---|---|
| | By | Date |
| | | |

## 4.13.12
## Sample Meeting / Conversation Record Form

Record of:

_____ Meeting
_____ Telephone Conversation

Date:____________________
Time: ______:______ AM / PM

Project: ____________________________
Proj. #:____________________________

Subject: ____________________________________________________________

Present / Calling:

Distribution:

Discussion:

Prepared By:

______________________________________

Date:_________________________________

**NOTICE:**
Failure to notify the writer of any necessary corrections to this Record constitutes acceptance of all information as it is represented.

- Clarify or hide real objectives
- Force quick decisions
- Restrict discussions or permit digressions
- Establish firm timetables or allow open ends
- Guide actions directly toward problem resolution in the shortest possible time

Regular job meetings have prearranged agendas in the form of "old business" and a fairly standard format for new items. Even there, however, the agendas of individual items can still be controlled.

2. *Action Rules.* The action rules that follow can be conducted as clear, distinct steps, as they appear in the case of larger, or special, issues. Even for simple issues, however, each step is still part of the process when approaching every issue at every regular or special meeting, no matter how formally or informally applied.

   a. *Confirm attendance prior to any meeting.* If specific individuals are required to resolve any issue, make certain they are present (see also Sections 4.13.6, 7, 8, and 14).
   b. *Think through the entire problem resolution process.* List each step, along with the corresponding people necessary to settle it. Catalog the specific actions necessary by each individual, along with the early and latest acceptable dates for their actions. This catalog will become your agenda.
   c. *Be sure that you are thoroughly prepared to discuss every issue and subissue.*

      - Have complete information.
      - Have your presentation package finalized.
      - List every conceivable option, along with the corresponding answer.
      - Practice; know all the bases and be able to cover them spontaneously.

   d. *Prior to the meeting, notify all expected participants of your specific agenda.* This will remove excuses on their parts that they do not have appropriate people available, have not secured required information, or are not otherwise prepared to discuss *and resolve* the issues.
   e. *Include the expectation of problem resolution in your agenda notification.*
   f. *Be aware of other people's efforts to control or divert your agenda.* Be prepared to force corrections.
   g. *Have good reasons for forcing compliance with your prearranged agenda.* Know and be believable with your own excuses for *not* discussing items that are not on the planned list:

      - Key or affected people are not available.
      - There is limited time available to cover the prearranged items; the new items must be tabled for another time.
      - Use "I don't know" if you need to.
      - You have every right to be unprepared for "surprises" (whether you're prepared or not…).
      - Do not discuss any item unless you are fully prepared to do so.

## 4.13.14 Sample Letter Confirming a Special Meeting *(page 4.66)*

The Sample Letter Confirming a Special Meeting is an example of a confirming letter designed to follow through on the recommendations of the previous sections. Specifically, it:

## 4.13.14
## Sample Letter Confirming a Special Meeting

Letterhead

(Date)

To:

RE: (Project)
(Company Project #)

SUBJ: Specific Issue(s)

Mr. (Ms.) ( ):

Confirming our conversation today,, a meeting will be held at my office on (Date) at (Time) to resolve the subject issue(s). The specific agenda is as follows:

1. (Primary Issue #1)
   a. (Sub Issue #1.1)

2. (Primary Issue #2)
   a. (Sub Issue #2.1)
   b. (Sub Issue #2.2)

As we discussed, please be sure that your Mr. (Name), Ms. (Name), and any other people necessary to completely resolve the issue(s) are present.

Thank you for your consideration.

Very truly yours,

COMPANY

Project Manager

cc: (List all those named in the letter)
(List all those definitely or potentially affected)
Jobsite
File: Vendor File:__________
CF

1. Confirms the meeting parameters of date, place, and time
2. Establishes the complete meeting agenda
3. Lists all those expected to attend
4. Notifies all expected attendees by copy of the letter

## 4.14 Securing Lien Waivers

### 4.14.1 General

Simply stated, a lien is a security interest in the particular real estate that has been improved. It is placed to secure payment for labor and material used in the property's improvements. The lien provides for the right to sell the property to which the lien attaches if the debt is not paid. A lien waiver is a short document executed to waive an individual's or a company's right to assert a lien. The right to assert a lien is not one recognized by common law, and is therefore defined by statute. Lien laws exist in every state and vary greatly in their terms.

This section does not discuss the intricacies of liens and lien laws. For that you must consult with a competent attorney to confirm the specific requirements and details of local law. Besides purely statutory rights and requirements, these intricacies will be further qualified by other material considerations such as:

- Type of property owner, whether state, federal, or municipal agency or private owner
- Presence or lack of payment and performance bonds
- Presence or lack of other available remedies for nonpayment under the terms of the general contract or subcontract

As a contracting professional, it is incumbent upon you to become intimately and completely familiar with all intricacies of liens and lien waivers for every geographic location in which you do business. They will profoundly affect daily operating decisions regarding treatment of Owners and Subcontractors for their appropriate and inappropriate actions regarding all project payments. Have your attorney provide you with the appropriate full and partial lien waiver forms to be used for each project.

This section, then, focuses on *securing* proper lien waivers from Subcontractors and Suppliers as a requirement for project payments—after those requirements have been correctly determined.

### 4.14.2 Full versus Partial Waivers of Lien

Since lien rights are a creature of statute and not of common law, they can be waived. The waivers themselves are necessary, both as a protection of the Company and as an inducement to the Owner to release additional payment.

Many general requisition forms and procedures to the Owner will require the delivery of appropriate lien waivers for all payments in prior periods as a condition of releasing the current payment. If a single Sub-Sub-Vendor is delinquent in the delivery of a single waiver, it can have the effect of locking up the entire general payment.

Attention must therefore be continually and sharply focused on maintaining *all* required waivers on an absolutely current basis.

*Partial Waivers of Lien* are documents executed by the payee to waive its right to assert a lien for an amount equal to the respective payment. This is usually a straightforward procedure accepted by most without much objection.

*Full Waivers of Lien* generally waive *all* rights to every lien on the property that is the subject of the agreement for work completed, *or yet to be completed.*

Although requiring a party to execute a full waiver may appear to be an extreme measure, in reality it may be a convenient mechanism that either can streamline contract procedures without actually compromising rights, or will keep dispute resolution options focused on the actual dispute resolution provisions defined in your subcontracts, for example, without allowing sharp attorneys to pervert the intention of your agreement.

**Example:** By statute, state-owned property in Connecticut cannot be longed. In place of lien rights, General Contractor payment and performance bonds are provided in order to preserve payment rights of Sub-Vendors.

Although there are no lien rights, some contracts may still actually require delivery of lien waivers as a condition of payment. In these cases, requiring a full waiver from all Sub-Vendors at the start of the project will relieve the logistical effort necessary to secure each waiver during every payment cycle.

**Example:** If you as a General Contractor have provided the Owner with a 100 percent payment bond, and your subcontract agreement provides for arbitration after a certain dispute-resolution procedure is to be followed, payment rights of the Subcontractor have been substantially protected. Leaving Subcontractor lien rights in place in these conditions only allows them to irritate the Owner and cause other legal problems that are expressly outside the dispute-resolution *intentions* of the subcontract. Requiring a Subcontractor to execute a full waiver puts the problem resolution back into the subcontract procedure and not onto the property.

### 4.14.3 Securing Sub-Vendor Lien Waivers

1. *Full Lien Waivers.* If the execution of a full lien waiver has been a condition of a subcontract or purchase order, have it executed at the time the subcontract is executed. If for any reason this has not been done, have the waiver executed upon delivery of the *first* progress payment; as a condition of payment.

   Refer to Section 4.7 for related discussion on contract execution and distribution procedures.

2. *Partial Lien Waivers*

   *a.* The *Sample Subcontract/Purchase Order Transmittal Form Letter* of Section 4.7 is the first written notification of the requirement after the subcontract or purchase order agreement itself.

   *b.* The *Sample Form Letter to Subcontractors Regarding Submittal Requirements* of Section 4.10.3 is the second written notification to provide the forms.

   *c.* In *every* case where it is at all practical (regardless of whether it is "inconvenient" for the payee) have an *authorized* representative of the Company pick up the periodic payments and execute the waiver as the condition of payment. If the waiver is not properly executed for whatever reason, the payment should not be released.

   *d.* In cases where it is not logistically practical for a payee to pick up a payment (such as a supplier across the country), conditions must be provided for that will allow for the next payment to be withheld until the correct waiver for previous payments has been delivered.

*e.* As a condition of payment it is the primary Sub-Vendor's responsibility to provide lien waivers for its Sub-Sub-Vendors for payment for all materials through all previous payments. If these waivers are not provided, the current payment to the primary Sub-Vendor must not be released.

*f.* Refer to Section 4.14.4 for the sample letter to induce compliance from your Sub-Vendors.

### 4.14.4 Sample Letter to Subcontractors/ Suppliers Regarding Failure to Provide Lien Waivers *(page 4.70)*

The Sample Letter to Subcontractors/Suppliers Regarding Failure to Provide Lien Waivers that follows is to be used in every case where there is an observed deficiency in any Sub-Vendor providing correct lien waivers for itself or for any of its Sub-Sub-Vendors. It is used to:

1. Summarize the specific waiver forms necessary
2. Notify that failure to provide the correct forms will result in delay in payment to all affected Vendors and Sub-Vendors
3. Notify the party that they will be held responsible for any effects on other parties resulting from the noncompliance

Refer to Section 4.15 for a form combining these notifications with those for Certified Payroll Reports as discussed there.

## 4.15 Securing Sub-Vendor Certified Payroll Reports

### 4.15.1 General

In many states and on federal projects, any project providing for the use of public funds will likely require the payment of prevailing wages under the provisions of the Davis-Bacon Act or other minimum-wage payment requirements.

On such projects, confirmation of the payment by Subcontractors of such prevailing wages will be done by way of Certified Payroll Reports. These are reporting forms on which the respective employer certifies to the Owner the actual amount of wages, taxes, and fringe benefits paid to the individual employees for hours worked on particular dates.

Section 4.15.2 displays an example of a Certified Payroll Report Form that is commonly used on state-funded projects. Check with your state or federal labor board for the correct forms to be used on the particular project as required.

This section, then, focuses on *securing* Certified Payroll Report Forms from each project Subcontractor on a current basis as a condition of payment.

### 4.15.2 Example Certified Payroll Report Form *(pages 4.72 and 4.73)*

An example of a Certified Payroll Report Form that is commonly used on state-funded projects follows. Check with your state or federal labor board for the correct forms to be used on a particular project as required.

## 4.14.4
## Sample Form Letter to Subcontractors/Suppliers Regarding Failure to Provide Lien Waivers

**Letterhead**

Date: ______________________________

To: ______________________________ Faxed: No: ____ Yes: ____
______________________________ Fax #: ____________________
______________________________

ATTN: ______________________________

RE: Project: ________________________
Company Project #: ____________________

SUBJ: Failure to Provide Subcontractor/Supplier Lien Waivers

Mr. (Ms.) ____________________________:

To date, we have not received properly executed lien waivers for:

____ Your Material Suppliers: ______________________________
______________________________
______________________________
All others as applicable.

____ Your Sub-subcontractors: ______________________________
______________________________
______________________________
All others as applicable.

These original documents in correct form and properly executed by those properly authorized to do so are to be delivered to this office by or before ____________________, 19______.

NO PAYMENT OTHERWISE DUE CAN BE RELEASED UNTIL THESE REQUIREMENTS HAVE BEEN MET.

Failure to comply may also affect the payments of all other subcontractors and suppliers for the period(s) represented, and may also dely the current general payment. In that event, you will be held responsible for all resulting costs and effects.

Thank you for your cooperation.

Very truly yours,
COMPANY

______________________________
Project Manager

cc: File: Vendor File________
CF

### 4.15.3 Payroll Liabilities on Construction Projects

On projects requiring prevailing wages, such wages are normally required for work performed at the site, but not for fabrication processes performed off-site.

If a Sub-Vendor fails to pay such prevailing wages to its employees, and a complaint is filed by any of its employees within the statutory period, the state or federal labor department will normally look for satisfaction of such payment of back wages and benefits in the following order:

1. The Sub-Vendor who has been confirmed to be in violation of the wage laws
2. The General Contractor
3. The payment bond

Because the payment bond is usually the last recourse, the project Owner will not normally have any liability. As such, you should not tolerate any efforts by any project Owner to withhold any payment because of an alleged violation of labor laws on bonded projects. Because prevailing wage projects are usually bonded, this should apply to all such projects.

Because of their recognized lack of liability, project Owners vary greatly in their attitudes toward policing wage payments. Some are relentless, while others will go through an entire project without ever asking for them.

The General Contractor, however, should be fully aware of the amount of extreme liability assumed if Sub-Vendors are allowed to violate wage laws, or if at least measures are not taken to confirm that the wages certified to be paid have actually been paid.

To help with such periodic confirmation, maintain efforts on all prevailing wage jobsites to continually discuss requirements with the individual workers. Follow up on any hint of a violation. Beyond these efforts, require first in your subcontract agreements, and then in practice, the delivery of the Subcontractor's actual weekly payroll ledger to support the Certified Payroll Report Forms being submitted.

Use the sample form letter of Section 4.15.4 to help in the enforcement of delivery of the proper Payroll Report Forms and payroll ledger, and *release no payments unless the requirements are completely fulfilled on a current basis.*

### 4.15.4 Sample Letter to Subcontractors Regarding Delivery of Certified Payroll Report Forms *(page 4.74)*

Whenever Certified Payroll Report Forms are required, they will be required to be submitted *weekly.* This is an express condition of payment. No payment can be released until the proper, complete forms have been delivered.

Use the Sample Letter to Subcontractors Regarding Delivery of Certified Payroll Report Forms to:

1. Identify the particular weeks of deficiency
2. Notify the offending party that failure to provide the correct forms is a breach of contract and will cause delay in payment until the condition is corrected

U.S. DEPARTMENT OF LABOR
WAGE AND HOUR DIVISION

# PAYROLL

**(For Contractor's Optional Use; See Instruction, Form WH–347 Inst.)**

Form Approved.
Budget Bureau No. 44–R1093

NAME OF CONTRACTOR ☐ OR SUBCONTRACTOR ☐ | ADDRESS

PAYROLL NO. | FOR WEEK ENDING | PROJECT AND LOCATION | PROJECT OR CONTRACT NO.

| (1) NAME, ADDRESS, AND SOCIAL SECURITY NUMBER OF EMPLOYEE | (2) NO. OF WITHHOLDING EXEMPTIONS | (3) WORK CLASSIFICATION | OT. OR ST. | (4) DAY AND DATE | | | | | | | (5) TOTAL HOURS | (6) RATE OF PAY | (7) GROSS AMOUNT EARNED | (8) DEDUCTIONS | | | | | | (9) NET WAGES PAID FOR WEEK |
|---|---|---|---|---|---|---|---|---|---|---|---|---|---|---|---|---|---|---|---|---|
| | | | | HOURS WORKED EACH DAY | | | | | | | | | | FICA | WITH-HOLDING TAX | | | OTHER | TOTAL DEDUCTIONS | |
| | | | O | | | | | | | | | | | | | | | | | |
| | | | S | | | | | | | | | | | | | | | | | |
| | | | O | | | | | | | | | | | | | | | | | |
| | | | S | | | | | | | | | | | | | | | | | |
| | | | O | | | | | | | | | | | | | | | | | |
| | | | S | | | | | | | | | | | | | | | | | |
| | | | O | | | | | | | | | | | | | | | | | |
| | | | S | | | | | | | | | | | | | | | | | |
| | | | O | | | | | | | | | | | | | | | | | |
| | | | S | | | | | | | | | | | | | | | | | |
| | | | O | | | | | | | | | | | | | | | | | |
| | | | S | | | | | | | | | | | | | | | | | |
| | | | O | | | | | | | | | | | | | | | | | |
| | | | S | | | | | | | | | | | | | | | | | |
| | | | O | | | | | | | | | | | | | | | | | |
| | | | S | | | | | | | | | | | | | | | | | |
| | | | O | | | | | | | | | | | | | | | | | |
| | | | S | | | | | | | | | | | | | | | | | |

FORM WH–347 (1/68) – FORMERLY SOL 184—*PURCHASE THIS FORM DIRECTLY FROM THE SUPT. OF DOCUMENTS*

Date ____________________

I, ____________________ (Name of signatory party), ____________________ (Title)

do hereby state:

(1) That I pay or supervise the payment of the persons employed by ____________________

____________________ (Contractor or subcontractor) on the ____________________ (Building or work)

____________________; that during the payroll period commencing on the ____________

day of ____________, 19____, and ending the ____ day of ____________, 19____, all persons employed on said project have been paid the full weekly wages earned, that no rebates have been or will be made either directly or indirectly to or on behalf of said

____________________ (Contractor or subcontractor) from the full

weekly wages earned by any person and that no deductions have been made either directly or indirectly from the full wages earned by any person, other than permissible deductions as defined in Regulations, Part 3 (29 CFR Subtitle A), issued by the Secretary of Labor under the Copeland Act, as amended (48 Stat. 948, 63 Stat. 108, 72 Stat. 967; 76 Stat. 357; 40 U.S.C. 276c), and described below:

____________________

____________________

____________________

____________________

(2) That any payrolls otherwise under this contract required to be submitted for the above period are correct and complete; that the wage rates for laborers or mechanics contained therein are not less than the applicable wage rates contained in any wage determination incorporated into the contract; that the classifications set forth therein for each laborer or mechanic conform with the work he performed.

(3) That any apprentices employed in the above period are duly registered in a bona fide apprenticeship program registered with a State apprenticeship agency recognized by the Bureau of Apprenticeship and Training, United States Department of Labor, or if no such recognized agency exists in a State, are registered with the Bureau of Apprenticeship and Training, United States Department of Labor.

(4) That:

(a) WHERE FRINGE BENEFITS ARE PAID TO APPROVED PLANS, FUNDS, OR PROGRAMS

☐ —In addition to the basic hourly wage rates paid to each laborer or mechanic listed in the above referenced payroll, payments of fringe benefits as listed in the contract have been or will be made to appropriate programs for the benefit of such employees, except as noted in Section 4(c) below.

(b) WHERE FRINGE BENEFITS ARE PAID IN CASH

☐ —Each laborer or mechanic listed in the above referenced payroll has been paid, as indicated on the payroll, an amount not less than the sum of the applicable basic hourly wage rate plus the amount of the required fringe benefits as listed in the contract, except as noted in Section 4(c) below.

(c) EXCEPTIONS

| EXCEPTION (CRAFT) | EXPLANATION |
|---|---|
| | |
| | |
| | |
| | |
| | |
| | |
| | |
| | |
| | |

REMARKS

| NAME AND TITLE | SIGNATURE |
|---|---|
| | |

THE WILFUL FALSIFICATION OF ANY OF THE ABOVE STATEMENTS MAY SUBJECT THE CONTRACTOR OR SUBCONTRACTOR TO CIVIL OR CRIMINAL PROSECUTION. SEE SECTION 1001 OF TITLE 18 AND SECTION 231 OF TITLE 31 OF THE UNITED STATES CODE.

## 4.15.4
## Sample Form Letter to Subcontractors Regarding Delivery of Certified Payroll Report Forms

**Letterhead**

Date: ______________________

To: ______________________ Faxed: No: ____ Yes: ____
______________________ Fax #: ______________
______________________

ATTN: ______________________

RE: Project: ______________________
Company Project #: ______________

SUBJ: Certified Payroll Reports

Mr. (Ms.) ______________________:

To date, we have not received properly completed Certified Payroll Reports for the following time periods:

______________________ ______________________
______________________ ______________________
______________________ ______________________
______________________ ______________________

These original documents in correct form and executed by those properly authorized to do so are to be delivered to this office by or before ______________, 19_____. Be sure that the weekly payroll ledgers accompany these submissions in order to support each respective Payroll Report Form.

NO PAYMENT OTHERWISE DUE CAN BE RELEASED UNTIL THESE REQUIREMENTS HAVE BEEN MET.

Failure to comply may also affect the payments of all other subcontractors and suppliers for the period(s) represented, and may also dely the current general payment. In that event, you will be held responsible for all resulting costs and effects.

Thank you for your cooperation.

Very truly yours,

COMPANY

______________________
Project Manager

cc: File: Vendor File________
Certified Payroll Reports
CF

3. Notify the party that they will be held responsible for any effects on other parties as a result of noncompliance

Refer to Section 4.14.4 regarding a similar procedure for securing lien waivers, and to Section 4.15.5 for a sample letter combining the forms for Certified Payroll Reports with those for lien waivers.

### 4.15.5 Sample Letter to Subcontractors Regarding Delivery of Lien Waiver Forms and Certified Payroll Report Forms *(page 4.76)*

## 4.16 Backcharges

### 4.16.1 General

A backcharge is a charge against a Subcontractor's or Supplier's account to cover the cost of having had to perform some portion of their responsibility for them by other means, or to correct some portion of their work. It is the contract mechanism that reimburses the Company for all expenses incurred, including but not limited to supervision (and other management efforts), overhead, and profit.

It is not the objective of this section to necessarily encourage Project Managers, Project Engineers, and Site Superintendents to pursue backcharges aggressively in every possible situation. Rather, it is here to:

1. Encourage project management to:
   - *a.* Keep adequate records and substantiation
   - *b.* Provide and maintain adequate notification
   - *c.* Be able to prepare an appropriate backcharge whenever necessary, and to support it properly
2. Establish and maintain project environments in which each Subcontractor and Supplier is aware of Company policies and procedures regarding backcharges. In so doing, Sub-Vendors who know of the Company's ability to do its work when necessary, support the charges, and their willingness to do so whenever necessary will have less of a tendency to allow matters to deteriorate to that level very often.

### 4.16.2 Conditions of the Backcharge

1. The Subcontractor or Supplier has failed to or refuses to perform the work in accordance with the requirements of its bid package in any respect. This applies to:
   - *a.* Physical work not completed at all
   - *b.* Completed work requiring correction
   - *c.* All administrative requirements of the bid package
2. The Subcontractor or Supplier was *notified* that the respective work must be completed or corrected by a particular date.
3. The party cannot or will not comply with the directive to complete or correct the work, or to do so by the time required.

## 4.15.5
## Sample Form Letter to Subcontractors
## Regarding Delivery of Lien Waiver Forms and Certified Payroll Report Forms

Letterhead

Date: ______________________________

To: ______________________________ Faxed: No: ____ Yes: ____
______________________________ Fax #: ____________________
______________________________

ATTN: ______________________________

RE: Project: ______________________________
Company Project #: ______________________

SUBJ: 1) Certified Payroll Reports 2) Subcontractor/Supplier lien Waivers

Mr. (Ms.) ____________________________:

To date, we have not received:

___ Properly completed Certified Payroll Reports for the following time periods:
__________________________________ __________________________________
__________________________________ __________________________________

___ Properly executed lien waivers for:
___ Your Sub-Subcontractors: ____________________________________________
____________________________________________
____________________________________________
All others as applicable.

___ Your Material Suppliers: ____________________________________________
____________________________________________
____________________________________________
All others as applicable.

These original documents in correct form and executed by those properly authorized to do so are to be delivered to this office by or before ____________________, 19______. NO PAYMENT OTHERWISE DUE CAN BE RELEASED UNTIL THESE REQUIREMENTS HAVE BEEN MET.

Failure to comply may also affect the payments of all other subcontractors and suppliers for the period(s) represented, and may also dely the current general payment. In that event, you will be held responsible for all resulting costs and effects.

Thank you for your cooperation.

Very truly yours,
COMPANY

____________________________________
Project Manager cc: File: Vendor File________, CF

### 4.16.3 Backcharge Procedure

1. Establish that the work (subject of the proposed backcharge) is required:

   *a.* To allow the job to progress
   *b.* To comply with any project requirement

2. Confirm *all* requirements of the contract documents:

   *a.* The general documents between the construction force and the Owner
   *b.* The specific subcontract or purchase order of the respective bid package being considered for the charge

3. Notify the party verbally (in a meeting, conversation, or telephone call). Advise:

   *a.* Of the *specific* work to be done
   *b.* Of the date by which the work is to be complete
   *c.* That if the work is not completed as directed, you will arrange to complete it by whatever means necessary, and charge the party for *all* associated costs.

   *Be prepared to actually move on this notice. Never threaten to do anything that you're not really prepared to do.*
4. Follow up with the Sample Backcharge Notice of Section 4.16.4. There *must* be such a written notice sent immediately.
5. When the work is not completed or the item not complied with by the deadline, *follow through* on your notice. Do not create any environment in which your notifications become viewed as idle gestures.

   *Immediately* arrange to complete the item by whatever reasonable means necessary. It is often useful to arrange to have the work completed by another Subcontractor, rather than with your own forces. This can go a long way later when the original Subcontractor argues about the size of the actual cost of the corrective work.
6. Take photographs before, during, and after all work. Refer to Section 5.16 for instructions on the proper procedure and identification of such photos.
7. Process the backcharge as a regular change order to the Sub-Vendor's contract. Be sure to include all costs associated with:

   *a.* Soliciting prices, coordination, and other direct efforts necessary to arrange for the work to be done
   *b.* Field coordination and supervision
   *c.* Prevailing or other appropriate wage rates
   *d.* Trucking and transportation charges
   *e.* Appropriate rates for overhead and profit

8. Use the Backcharge Summary Log of Section 4.16.7 to keep a running record of smaller incidents that can be more conveniently consolidated in the completed change order after the total costs reach an appropriate magnitude.

### 4.16.4 Use of Backcharge Notice Form Letter

The Backcharge Notice Form Letter is used to follow through with the notification requirement of the previous discussion. Specifically, it:

1. Establishes the true date and time of notification as of the moment of conversation (*not* as of the receipt of the written notice)

2. Delivers the written notice as soon as possible (by fax and mail), thereby helping to confirm the urgent nature of the notice
3. Concisely but completely describes the work required, along with the subsequent work that is being immediately affected by the work in question
4. Gives the firm deadline by which acceptable performance must be achieved
5. Notifies of your intent to complete the work and charge the Sub-Vendor's account

### 4.16.5 Sample Backcharge Notice Form Letter—Completed Example *(page 4.79)*

### 4.16.6 Sample Backcharge Notice Form Letter—Blank Form *(page 4.80)*

### 4.16.7 Use of Backcharge Summary Log

The Backcharge Summary Log is a form prepared for the individual Sub-Vendor upon the initial incident creating any potential for the first backcharge. It will be prepared along with the first Backcharge Notice as described in previous sections, and will remain a part of the permanent Sub-Vendor File.

Large backcharge items will justify the immediate preparation of the appropriate deduct change order to the Sub-Vendor's subcontract or purchase order. Small backcharge items can be listed on the Backcharge Summary Log as they occur, and consolidated at any convenient time into a single or a few such change orders.

The log can be kept in the respective Sub-Vendor's General File Folder (see Section 2.4) or behind the Subcontractor Summary Form of Section 4.4 in the Subcontractor Summary and Telephone Log book.

### 4.16.8 Sample Backcharge Summary Log—Completed Example *(page 4.81)*

### 4.16.9 Sample Backcharge Summary Log—Blank Form *(page 4.82)*

### 4.16.10 Site Cleanup—A Special Case

General housekeeping and site cleanup are areas of chronic problems with Subcontractors. If the site is not policed with extreme diligence by the Site Superintendent, it quickly becomes a mess. When that is allowed to happen, each Contractor loses a corresponding amount of initiative to keep its own work clean. The process will continue to degenerate until the site is such a mess that it is everyone's fault—but it will be "no one's fault" because the mess will be a combination of everybody's materials.

## 4.16.4
## Sample Backcharge Notice
## (Completed Example)

**Letterhead**

Date: AUGUST 20, 1994

To: CRACK PLASTER CO., INC.
519 HEART TPK
CANTON, CT 06902

Confirmation of Fax:
Fax #: 984-9920

ATTN: PETER CRACK

RE: Project: RAINBOW CHILD CARE
Proj. #: 9442

SUBJ: CONTINUING SLIPPAGE IN PLASTER WORK COMPLETION SCHEDULE

Mr. ~~/Ms.~~ CRACK:

Confirming our conversation at 9:10 (AM) / PM this date, please:
CORRECT THE PLASTER PATCHING AT ROOMS 203, 204, & 205
IN ORDER TO ALLOW PAINTING TO CONTINUE.

This work must be completed by 5:00 AM / (PM) on AUGUST 22, 1994.

If the work is not completed by then, we will make arrangements to complete the work for you, and backcharge your company for all costs incurred, including mobilization, supervision, overhead and profit.

Very truly yours,

COMPANY

David Veltko
PROJECT ENGINEER

cc: Accounts Payable
File: Vendor PLASTER
CF

# 4.16.5
# Sample Backcharge Notice

**Letterhead**

Date: ______________________________

To: ______________________________
______________________________
______________________________

Confirmation of Fax:
Fax #: ____________________

ATTN: ______________________________

RE: Project: ________________________
Proj. #: ________________________

SUBJ: ____________________________________________________________

Mr. / Ms. ____________________________:

Confirming our conversation at _____:_____ AM / PM this date, please:

____________________________________________________________
____________________________________________________________
____________________________________________________________
____________________________________________________________
____________________________________________________________
____________________________________________________________
____________________________________________________________
____________________________________________________________
____________________________________________________________

This work must be completed by _____:_____ AM / PM on ____________________, 19_____.

If the work is not completed by then, we will make arrangements to complete the work for you, and backcharge your company for all costs incurred, including mobilization, supervision, overhead and profit.

Very truly yours,

COMPANY

______________________________
______________________________

cc: Accounts Payable
File: Vendor__________
CF

# 4.16.8
# Sample Backcharge Summary Log
# (Completed Example)

## Backcharge Log

Project: NEWTOWN COMMUNITY CNTR
Proj. #: 9900
Subcontractor / Supplier: BRICKYARD MASONRY CONTR, INC.

| Date Notification Sent | Work Req'd To Be Complete By | Actual Date Completed | Completed By (Name) | Total Cost | OH & P | TOTAL |
|---|---|---|---|---|---|---|
| 9-9-94 | 9-11-94 | 9-12-94 | C-HILL | 940.00 | 142.00 | 1,082.00 |
| 10-10-94 | 10-12-94 | 10-16-94 | MCCI | 220.00 | 33.00 | 253.00 |
| 12-2-94 | 12-4-94 | 12-5-94 | C-HILL | 350.00 | 53.00 | 383.00 |
| | | | | | | |

# 4.16.9
# Sample Backcharge Summary Log

## Backcharge Log

Project: ______________________________ Subcontractor / Supplier:

Proj. #: ______________________________ ______________________________

| Date Notification Sent | Work Req'd To Be Complete By | Actual Date Completed | Completed By (Name) | Total Cost | OH & P | TOTAL |
|---|---|---|---|---|---|---|
| | | | | | | |
| | | | | | | |
| | | | | | | |
| | | | | | | |
| | | | | | | |
| | | | | | | |
| | | | | | | |
| | | | | | | |
| | | | | | | |
| | | | | | | |
| | | | | | | |
| | | | | | | |
| | | | | | | |
| | | | | | | |
| | | | | | | |
| | | | | | | |
| | | | | | | |
| | | | | | | |
| | | | | | | |
| | | | | | | |
| | | | | | | |
| | | | | | | |
| | | | | | | |
| | | | | | | |
| | | | | | | |
| | | | | | | |
| | | | | | | |
| | | | | | | |
| | | | | | | |
| | | | | | | |
| | | | | | | |
| | | | | | | |
| | | | | | | |
| | | | | | | |
| | | | | | | |
| | | | | | | |
| | | | | | | |
| | | | | | | |
| | | | | | | |
| | | | | | | |
| | | | | | | |
| | | | | | | |
| | | | | | | |
| | | | | | | |
| | | | | | | |
| | | | | | | |

Cleanup should be a primary consideration at every Subcontractor's meeting and throughout every workday. No violation is to be tolerated.

Use the Backcharge Notice procedure as described as a *weekly routine* to ensure that by every Friday each Subcontractor has arranged to clean up all materials and debris. If not, you must keep making arrangements to clean up the site for them.

In these cases it is crucial that the associated backcharges be processed *immediately.* This will help give the message to all Vendors that you're absolutely serious. In most cases, Subcontractors will then fall in line with their cleanup responsibilities—at least until the next time.

### 4.16.11 Sample Letter to Subcontractors Regarding Disregard for Finishes *(page 4.84)*

The Sample Letter to Subcontractors Regarding Disregard for Finishes is a special notification to be sent to all subcontractors at the appropriate stage of project completion when finishes are proceeding.

Although you should have the right to expect that trades people would automatically respect finish construction as it is being installed, this unfortunately cannot be relied upon in too many cases.

The letter first calls attention to the stage of completion of the project and to your expectation of proper consideration. After that it will serve as your notice of backcharges if such negligence and damage should occur.

## 4.17 The Punchlist: Expediting Final Completion

### 4.17.1 General

The "punchlist" should be on everyone's mind throughout the completion of any and every item of work. If anything is observed during the course of construction that you know is not satisfactory as finished work, complete it now—don't wait for it to find its way onto an official distributed list and possibly have some cost assigned to it.

Let the Owner and Architect observe you taking this approach. It will go a long way in building their confidence in you and in the final completed product.

As the project nears substantial completion, prepare your own punchlist, and prosecute its completion with the various Subcontractors. Do this in an effort to keep the official punchlist small.

When the project is sufficiently ready, typically at the point of or immediately after substantial completion, notify the Architect that the project is ready for the punchlist inspection.

Upon receipt of the official punchlist, proceed as directed in the remainder of this section both to confirm to the Owner that you are proceeding with the punchlist completion, and to expedite the completion itself with the various Subcontractors.

## 4.16.11
## Sample Letter to Subcontractors Regarding Disregard for Finishes

| Letterhead |
| --- |

(Date)

To: (List all Project Subcontractors)

RE: (Project)
(Company Project #)

SUBJ: Care and Regard for Finishes

Gentlemen:

As the project nears completion, finish products are being installed and applied daily. At this time, negligence and disregard for the work of other trades will not be tolerated. Cleaning made necessary and/or damage caused to any finish work will be corrected at the expense of the responsible party(ies). Backcharge costs will include mobilization, preparation, protection, direct costs, supervision, overhead, and profit. Any person guilty of repeated and/or serious violation will be banned from the site.

We trust that everyone will demonstrate due consideration for the work of others, and hope there will be no need for these measures.

Thank you for your cooperation.

Very truly yours,

COMPANY

Project Manager

cc: Owner
Architect
Jobsite
File: Backcharge File
Vendor File:__________
CF

### 4.17.2 Sample Letter to the Architect Regarding Substantial Completion and Punchlist Review *(page 4.86)*

When the project is ready for inspection, use the Sample Letter to the Architect Regarding Punchlist Review to notify the architect that:

1. The project is substantially complete.
2. The punchlist should be prepared as soon as possible.
3. Every effort must be made to ensure that the punchlist is complete.

We'd like to see *one* punchlist. It is hoped that the construction force will then be mobilized to complete the punchlist work—we don't want to keep repeating the effort for multiple lists.

The letter will give you at least some basis to call foul if second and third punchlists appear after the work of the first one has been completed.

### 4.17.3 Punchlist Review and Distribution Procedure

1. Immediately upon receipt of the official punchlist, it should be reviewed to confirm that each item on it is legitimate; that the design professionals and/or the Owner are not attempting to make the construction force responsible for items of work that are not its fault. Items of wall repair and paint touch-up, for example, commonly fall into this category if the punchlist is allowed to be made after all or a portion of the project has been occupied.
2. Immediately advise the Architect of any item that in your opinion is in question. Remember, however, the "GC as Conduit" principle (Section 3.1.2) and the Pass-Through Clause (Section 3.3.9). Even though you may for the moment disagree with the item(s), they must still be forwarded with the current direction to the respective Subcontractor(s). You may after all discover that they'll get done without argument.
3. Mark each item on the punchlist to identify the specific party responsible to correct or complete it. It is often useful to assign a number to each Subcontractor, with a key written directly on the first page of the punchlist.
4. Review the list to determine that all administrative requirements are included, such as:

   *a.* Guarantees and warranties
   *b.* Attic stock
   *c.* Maintenance manuals
   *d.* As-Built Drawings
   *e.* System balancing reports
   *f.* System instruction
   *g.* Any other special requirements that will be necessary to close out the project

   If they have not been listed, add them yourself.

   Note that it will be of great advantage to you if you made arrangements throughout the project to secure these kinds of items well in advance of the punchlist. They typically are difficult and time-consuming to prepare and may be disproportionate in cost if you actually have to wind up doing them yourself.

   Having them completed in advance of the punchlist will also help to avoid any "hostage" situation if you have to arrange to finish items for any noncomplying Subcontractors and backcharge them for costs incurred.

# 4.17.2
# Sample Letter to the Architect
# Regarding Substantial Completion and Punchlist Review

| Letterhead |
|---|

(Date)

To: (Architect)

RE: (Project)
(Company Project #)

SUBJ: Substantial Completion and Punchlist Review

Mr. (Ms.) :

As of (Date), the project is substantially complete.

Please inspect the site as soon a possible to determine your punchlist of any items that you feel are necessary to correct prior to final completion.

It will be greatly appreciated if every effort is made to prepare as complete a punchlist as possible; it will be in everyone's best interest if we can all deal with a single list. This, as you know, will facilitate efficient completion of all items as expeditiously as possible.

Thank you for your consideration.

Very truly yours,

COMPANY

Project Manager

cc: Owner
Jobsite
File: Substantial Completion
Punchlist
CF

5. Transmit the punchlist to each Subcontractor, requiring completion of all items by a particular date. Indicate the specific problems resulting from noncompletion, and that other measures will be taken to complete the work by other means if necessary. Use the Sample Punchlist Notification Form Letters of the following sections to help in this procedure.
6. Refer to Section 4.18 for procedures with regard to securing appropriate guarantees and warranties.
7. After exhausting efforts with any original Subcontractor, consider moving ahead with your notice to complete the work by other means. Proceed carefully and completely, in a manner consistent with the discussion of Section 4.16, *Backcharges*.

### 4.17.4 Sample Punchlist Notification Form Letter #1 *(page 4.88)*

Use the Sample Punchlist Notification Form Letter #1 to:

1. Transmit the punchlist to each responsible party
2. Notify each party of the date by which all items must be completed
3. Notify each party that failure to comply will directly affect all project payments

Consider using the Sample Punchlist Notification Form Letter #2 (Section 4.17.5) as a stronger initial approach. Use Letter #3 (Section 4.17.6) to follow up on Letter #1 with those Subcontractors who have not complied with your direction.

### 4.17.5 Sample Punchlist Notification Form Letter #2 *(page 4.89)*

The Sample Punchlist Notification Form Letter #2 is a stronger version of Letter #1. It goes on to notify that failure to complete the work as directed may result in the Company making immediate arrangements to complete the work by whatever means available, and backcharge the delinquent party for all costs incurred. Consider further changing "may" to "will" to make the letter even stronger.

Note that it advises that the work will be completed by the "most expedient" means available. This is the notification that relieves later criticism of high cost to complete the work. Remember, it is *you* being forced into the position; not the Subcontractor.

Be prepared to actually follow through on the notice, however, if you decide to take this approach. Follow the considerations of Section 4.16, *Backcharges,* for related procedures.

### 4.17.6 Sample Punchlist Notification Form Letter #3 *(page 4.90)*

The Sample Punchlist Notification Form Letter #3 can be used as a second notification to follow up on Letter #1 for those Contractors failing to complete their work by the specified date.

## 4.17.4
## Sample Punchlist Notification Form Letter #1

**Letterhead**

Date: ______________________

To : ________________________________
________________________________
________________________________

CERTIFIED MAIL
RETURN RECEIPT REQUESTED

ATTN: ________________________________

RE: Project:________________________
Project #:______________________

SUBJ: Punchlist Completion

Mr. (Ms.)__________________________:

Attached is the Architect's Punchlist for the project.

All items pertaining to your work must be complete by or before ______________, 19_____. This is necessary in order to avoid delay in final completion and resulting delays in retainage release and final payment.

After you have completed your items, notify this office so that final inspection can be made.

In addition, immediately upon completion of all your items, return a copy of the Punchlist marked to indicate the dates that the respective items had been completed.

Thank you for your cooperation.

Very truly yours,

COMPANY

________________________________
Project Manager

cc: Jobsite
Vendor File:_________
Punchlist File
CF

# 4.17.5
# Sample Punchlist Notification Form Letter #2

| Letterhead |
|---|

Date: ______________________

To : ________________________________
________________________________
________________________________

CERTIFIED MAIL
RETURN RECEIPT REQUESTED

ATTN: ________________________________

RE: Project:________________________
Project #:________________________

SUBJ: Punchlist Completion

Mr. (Ms.)___________________________:

Attached is the Architect's Punchlist for the project.

All items pertaining to your work must be complete by or before _______________, 19_____. This is necessary in order to avoid delay in final completion and resulting delays in retainage release and final payment.

If your work is not complete by that date, other arrangements will be made to complete the work by whatever means available that are most expedient. In that event, you will be backcharged for all costs incurred, including mobilization, preparation, protection, direct costs, supervision, overhead, and profit. We accordingly will appreciate your efforts to avoid these possibilities.

After you have completed your items, notify this office so that final inspection can be made.

In addition, immediately upon completion of all your items, return a copy of the Punchlist marked to indicate the dates that the respective items had been completed.

Thank you for your cooperation.

Very truly yours,

COMPANY

______________________________________
Project Manager

cc: Jobsite
Vendor File:_________
Punchlist File
CF

## 4.17.6
## Sample Punchlist Notification Form Letter #3

**Letterhead**

Date: ______________________

To : ________________________________
________________________________
________________________________

ATTN: ________________________________

Confirmation of Fax
Fax#:____________________

CERTIFIED MAIL
RETURN RECEIPT REQUESTED

RE: Project:_________________________
Project #:_________________________

SUBJ: Punchlist Completion

Mr. (Ms.)__________________________:

You have been directed on a number of occasions to complete your punchlist work, but to date items remain to be completed.

As your final notice, be advised that these items must be absolutely complete by or before ______________________, 19______. After that date, arrangements will be made to complete your work by whatever means available that are most expedient. In that event, you will be backcharged for all costs incurred, including mobilization, preparation, protection, direct costs, supervision, overhead, and profit.

Please confirm to this office today whether you will complete your work as directed, or not.

Very truly yours,

COMPANY

______________________________________
Project Manager

cc: Jobsite
Vendor File:_________
Punchlist File
CF

Letter #1 was a softer initial notification approach. Letter #3 raises the severity of the communication by giving one last and final real deadline, after which costs will start to accumulate for the offender.

### 4.17.7 Sample Notice of Supervisory Costs for Late Final Completion *(page 4.92)*

The Sample Notice of Supervisory Costs for Late Final Completion can be sent to all Subcontractors who for any reason have not completed contract, punchlist, or guarantee work by the dates required. Consider using it as an interim notification to, in a sense, keep the wheel squeaking, as another prod to keep Subcontractors moving to complete their work, and as another notification that will help the Company recover at least some of the real costs of late completion.

## 4.18 Securing Subcontractor/Supplier Guarantees and Warranties

### 4.18.1 General

The process of securing Subcontractor and Supplier guarantees and warranties is very similar and closely related to that for the punchlist. In fact, these items will actually be part of the punchlist if they have not been secured by that time.

They are, however, a special case because of the nature of the items and the difficulty in providing acceptable documents to the Owner if they are not, for whatever reason, forthcoming from the original Sub-Vendor. If the acceptable documents are not provided, it can have a profound effect on the release of final payment, and even on retainage reduction. It is therefore incumbent upon the General Contractor to take a firm, direct approach to securing these documents as expeditiously as possible, and to take strong measures in the face of noncompliance.

The process described in this section effectively parallels that for the punchlist. Refer also to Section 3.13, *Guarantees and Warranties,* for important related discussion.

### 4.18.2 Procedure

1. As the project nears substantial completion, prepare your own list of required documents.
   *a.* Don't wait for the Architect or Owner representative to formally advise you of the need to provide general and specific guarantees and warranties.
   *b.* If a list of required guarantees and warranties is given in the supplementary general conditions, do not accept it as complete. Research every specification section yourself to confirm that the list is in fact complete. This will go a long way toward oversight of any item that can later affect final payment.
2. Transmit the requirement to each Vendor with any responsibility for providing specified guarantees and warranties. Use the *Sample Request for Guarantees/Warranties Form Letter* of Section 4.18.3 to help here.

## 4.17.7
## Sample Notice of Supervisory Costs for Late Final Completion

| Letterhead |
|---|

(Date)

To : (List All Project Subcontractors) (List all respective Fax #'s)

RE: (Project Description)
(Company Project #)

SUBJ: Supervisory Costs for Late Final Completion

Gentlemen:

As you are aware, the original Substantial Completion Date for this project was (Date). More than sufficient time has since been allowed for all trades to complete each's respective punchlist.

Be advised that after (Date), no (Company) supervisory personnel will be on the jobsite on any regular basis. Any of your contractual work progressing for any reason beyond that date will make it necessary for (Company) management and supervisory personnel to be at the site while your work is being performed. In that event, you will be backcharged accordingly.

Very truly yours,

COMPANY

Project Manager

cc: Jobsite
File: Punchlist
Backcharge
CF

3. Follow up with "second" and "third" requests if necessary. Write additional notes on the original request letter to stress the urgency, then photocopy and send out. Refer to the *Sample Guarantee / Warranty "Second Request" Letter* of Section 4.18.4 as an example.
4. If there still is no acceptable response, contact the President of the Company, or as high an official as you can reach. Advise him or her of the effort spent, time passed, and potential and real effects on the project and related payments. Be sure to send letters confirming every conversation.
5. When all other efforts have failed, consider using the *Sample Final Notice to Sub-Vendors to Provide Guarantees / Warranties* of Section 4.18.5. Be prepared to actually follow through on the notice—and follow through immediately on the prescribed date.

### 4.18.3 Sample Request for Guarantees/ Warranties Form Letter #1 *(page 4.94)*

The Sample Request for Guarantees/Warranties Form Letter #1 is another example of a similar form letter provided in Section 3.13.6. It is the first written notification to the Subcontractor or Supplier to provide all proper guarantees and warrantees. Specifically, it:

1. Advises that proper form and language are required
2. Specifies the date when the guarantee or warranty is to take effect
3. Informs payments will be affected until the proper submission is received

Fax and mail the letter.

### 4.18.4 Sample Guarantee/Warranty "Second Request" Letter *(page 4.95)*

The Sample Guarantee/Warranty "Second Request" Letter is to be sent out immediately on the date specified in Letter #1 to all those who have not submitted proper guarantees or warranties. Specifically, it:

1. Calls attention to the first notice, notice date, and originally required submission date
2. Notifies that effects on the project have already occurred for which the party will be held responsible
3. Notifies that the party will be held responsible for any additional effects resulting from failure to comply in a timely manner

### 4.18.5 Sample Final Notice to Sub-Vendors to Provide Guarantees/Warranties *(page 4.96)*

The Sample Final Notice to Sub-Vendors to Provide Guarantees/Warranties is to be sent to any Sub-Vendor who, despite every effort, simply fails to provide any guarantees or warranties, or fails to provide them in proper form or content. Specifically, it notifies the offender that:

1. The Company is now making other arrangements to provide the documents
2. An appropriate amount of money will be withheld from the Sub-Vendor's account to cover the cost of possible guarantee work

## 4.18.3
## Sample Request for Guarantees/Warranties Form Letter #1

| Letterhead |
|---|

Date)

To: (List All Project Subcontractors) (List All Respective Fax #'s)

RE: (Project)
(Company Project #)

SUBJ: Delivery of Guarantees and Warranties

Gentlemen:

It is your responsibility to submit guarantees and warranties in proper form. These must be delivered to this office by or before (Date). Please review your specifications to confirm specific content and language as detailed therein, and comply in every respect.

Release of retainage is contingent upon your compliance with these requirements.

If you have any questions, please contact me immediately.

Very truly yours,

COMPANY

Project Manager

cc: Jobsite
File: Guarantees/Warranties
CF

## 4.18.4
## Sample Guarantee/Warranty "Second Request" Letter

**Letterhead**

Date: ______________________

To : ________________________________
________________________________
________________________________

ATTN: ________________________________

Confirmation of Fax
Fax#:____________________

CERTIFIED MAIL
RETURN RECEIPT REQUESTED

RE: Project:__________________________
Project #:_________________________

SUBJ: Failure to Provide Proper Guarantees/Warranties

Mr. (Ms.)__________________________:

On _____________________, 19____, you were requested to provide all appropriate guarantees and warranties in accordance with the terms of your contract. To date:

____ We have not received them.

____ The documents have been submitted in unacceptable form.

This delay has already affected the current progress payment and any retainage reduction. It is also requiring excessive efforts to coordinate and enforce compliance by your company. You will be held responsible for these effects.

You are accordingly directed to provide the required guarantees/warranties in their proper form to this office by or before _____________________, 19____. Failure to comply will result in delay to the general requisition, payment to you, and payments to all other project vendors. In this event, you will be held responsible for all such effects.

Thank you for your cooperation.

Very truly yours,

COMPANY

______________________________________
Project Manager

cc: Jobsite
Vendor File:_________
Guarantees/Warranties File
CF

# 4.18.5
# Sample Final Notice to Sub-Vendors to Provide Guarantees/Warranties

**Letterhead**

Date: ____________________

To : ______________________________
______________________________
______________________________

ATTN: ______________________________

Confirmation of Fax
Fax#:__________________

CERTIFIED MAIL
RETURN RECEIPT REQUESTED

RE: Project:______________________
Project #:______________________

SUBJ: Failure to Provide Proper Guarantees/Warranties

Mr.(Ms.)________________________:

You were requested on a number of occasions to provide all appropriate guarantees and warranties in accordance with the terms of your contract. To date, you continue to fail to comply.

Accordingly, we are now making arrangements to provide the required guarantees and warranties to the Owner on your behalf in order to allow project closeout.

Until your proper guarantees/warranites are received by this office and accepted by the Owner, an appropriate amount of money as determined in our discretion will be withheld from your current payment as security to cover the cost of possible guarantee/warranty work. We estimate this value at $__________________.

In addition, your contract will be backcharged for all costs related to the excessive measures necessary to resolve this issue, including attorney's fees, administrative time, overhead, and profit.

Very truly yours,

COMPANY

______________________________
Project Manager

cc: Jobsite
Vendor File:________
Guarantees/Warranties File
CF

3. The Subcontractor will be backcharged for all the time, effort, and expense involved (since the initial effort)

It is a good idea to consult with the Owner and/or the Architect before proceeding. Advise them of the trouble experienced with the particular Sub-Vendor and try to elicit a confirmation that they will actually accept such alternative guaranties or warranties without other legal maneuvering. If they are not agreeable, consult with your attorney to determine *your* rights under the general contract to provide such alternate guaranties and warranties.

## 4.19 Sub-Vendor Performance Evaluation

### 4.19.1 Use of Sub-Vendor Performance Evaluation Form

The Sub-Vendor Performance Evaluation Form is to be completed by those with project manager, project engineering, and superintendent responsibilities—those who were closest to the performance itself.

It will be placed in the Sub-Vendor's permanent file to assist purchasing efforts in the future, and it can even be consolidated in any narratives or completion reports to be prepared at the end of the project.

Be sure that the form is completed for every Sub-Vendor on every project.

### 4.19.2 Sample Sub-Vendor Performance Evaluation Form—Completed Example *(page 4.98)*

### 4.19.3 Sample Sub-Vendor Performance Evaluation—Blank Form *(page 4.99)*

## 4.20 Project Closeout Checklist.*39*

The Project Closeout Checklist is to be considered as the project is nearing its completion. Hopefully, by the time the project is in this stage, many of the items listed will be either complete or in process.

The checklist summarizes the items most typical of conventional construction projects of any size that need to be completed, submitted, turned over, etc., in order to complete all detailed requirements of each bid package. Sections of it can therefore be conceptually repeated for each bid package, with satisfaction of the checklist itself considered to be the consolidation of the respective requirements for each bid package.

It is important, however, to acknowledge that such requirements are being "supplemented" almost daily—to subtle or dramatic degrees. It is therefore crucial that the specifications and any procedure manual for a specific project be studied in

## 4.19.2
## Sample Sub-Vendor Performance Evaluation Report (Completed Example)

Date: AUGUST 16, 1994
Project: HIGHTOWER APTS
Project #: 9990
Owner: PAUL HIGHTOWER
Architect: GEORGE ARCH
Location: 555 HIGHTOP PL.
Project Manager: MARK LEONARDO
Project Engineer: DAVE JEFFKO
Superintendent(s): PAUL SMALL

X Subcontractor ___ Supplier
Co. Name: STRAIT STEEL CORP
Address: 950 HIGHLAND PKWY
NEWTOWN, CT 06424
Contact: JOHN STRAIT
Phone: (203) 555-1652 Fax: (203) 555-1660

Scope of Work: STRUCTURAL STEEL
Subcontract/P.O. Value: $96,000.-
Start: Req'd: 4-20-94 Actual: 4-27-94
Comp.: Req'd: 5-1-94 Actual: 5-1-94

Project Type:
___ State ___ Federal ___ Municipal ___ Private

| | PERFORMANCE | EVALUATION | | | | |
|---|---|---|---|---|---|---|
| | | Excellent | | Good | | Poor |
| 1. | Understanding of Subcontract/P.O. | | X | | | |
| 2. | Understanding of Contract Documents | | X | | | |
| 3. | Timely preparation of Approval Submittals | X | | | | |
| 4. | Complete preparation of Approval Submittals | | X | | | |
| 5. | Understanding of/adherence to schedules | | X | | | |
| 6. | Record of material delivery schedules | X | | | | |
| 7. | Quality of communications | | | X | | |
| 8. | Timeliness of communications | | | X | | |
| 9. | Attendance at job meetings | | | | X | |
| 10. | Cooperation; willingness to resolve problems | | X | | | |
| 11. | Adequate management | | X | | | |
| 12. | Adequate number and ability of administrative staff | | X | | | |
| 13. | Ability & willingness to comply with project procedures | | | | X | |
| 14. | Timliness/completeness of Certified Payroll Reports | | | | X | |
| 15. | Timliness/completeness of lien waivers & releases | | | | X | |
| 16. | Worker payment record | | X | | | |
| 17. | Subcontractor payment record | | | N/A | | |
| 18. | Supplier payment record | | X | | | |
| 19. | Jobsite cleanup record | | | X | | |
| 20. | Ability to minimize punchlist | | X | | | |
| 21. | Overall level of workmanship | | X | | | |
| 22. | Submission of Guarantees & Warranties | | | X | | |
| 23. | Preparation/submission of As-Built documents | | | X | | |
| 24. | Proper submission of all project closeout documents | | | | X | |
| 25. | Safety compliances and safety performance | | X | | | |
| 26. | EEO & MBE employment compliances | | | X | | |
| 27. | Avoidance of claims and claims posturing | X | | | | |
| 28. | Avoidance of need for backcharges | X | | | | |
| 29. | TOTAL | 4 | 12 | 6 | 5 | 0 |

Describe the most significant good and/or poor performances: KNOWS TRADE & WORKMANSHIP WELL. CAN USE IMPROVEMENT IN TIMLINESS OF SUBMISSION OF OVERALL PROJECT DOCUMENTATION - BUT IT EVENTUALLY GETS SUBMITTED CORRECTLY. A PLEASURE TO WORK WITH ON-SITE

Prepared By: Mark Leonardo Date: 8/16/94

## 4.19.3
## Sample Sub-Vendor Performance Evaluation Report

Date: ______________________________ ___ Subcontractor ___ Supplier
Project: ______________________________ Co. Name: ______________________________
Project #: ______________________________ Address: ______________________________
Owner: ______________________________ ______________________________
Architect: ______________________________ Contact: ______________________________
Location: ______________________________ Phone: ______________ Fax: ______________
Project Manager: ______________________________
Project Engineer: ______________________________ Scope of Work: ______________________________
Superintendent(s): ______________________________ Subcontract/P.O. Value: ______________________________
______________________________ Start: Req'd: __________ Actual: __________
Comp.: Req'd: __________ Actual: __________

Project Type:
___ State ___ Federal ___ Municipal ___ Private

| | PERFORMANCE | EVALUATION | | | | |
|---|---|---|---|---|---|---|
| | | Excellent | | Good | | Poor |
| 1. | Understanding of Subcontract/P.O. | ___ | ___ | ___ | ___ | ___ |
| 2. | Understanding of Contract Documents | ___ | ___ | ___ | ___ | ___ |
| 3. | Timely preparation of Approval Submittals | ___ | ___ | ___ | ___ | ___ |
| 4. | Complete preparation of Approval Submittals | ___ | ___ | ___ | ___ | ___ |
| 5. | Understanding of/adherence to schedules | ___ | ___ | ___ | ___ | ___ |
| 6. | Record of material delivery schedules | ___ | ___ | ___ | ___ | ___ |
| 7. | Quality of communications | ___ | ___ | ___ | ___ | ___ |
| 8. | Timeliness of communications | ___ | ___ | ___ | ___ | ___ |
| 9. | Attendance at job meetings | ___ | ___ | ___ | ___ | ___ |
| 10. | Cooperation; willingness to resolve problems | ___ | ___ | ___ | ___ | ___ |
| 11. | Adequate management | ___ | ___ | ___ | ___ | ___ |
| 12. | Adequate number and ability of administrative staff | ___ | ___ | ___ | ___ | ___ |
| 13. | Ability & willingness to comply with project procedures | ___ | ___ | ___ | ___ | ___ |
| 14. | Timliness/completeness of Certified Payroll Reports | ___ | ___ | ___ | ___ | ___ |
| 15. | Timliness/completeness of lien waivers & releases | ___ | ___ | ___ | ___ | ___ |
| 16. | Worker payment record | ___ | ___ | ___ | ___ | ___ |
| 17. | Subcontractor payment record | ___ | ___ | ___ | ___ | ___ |
| 18. | Supplier payment record | ___ | ___ | ___ | ___ | ___ |
| 19. | Jobsite cleanup record | ___ | ___ | ___ | ___ | ___ |
| 20. | Ability to minimize punchlist | ___ | ___ | ___ | ___ | ___ |
| 21. | Overall level of workmanship | ___ | ___ | ___ | ___ | ___ |
| 22. | Submission of Guarantees & Warranties | ___ | ___ | ___ | ___ | ___ |
| 23. | Preparation/submission of As-Built documents | ___ | ___ | ___ | ___ | ___ |
| 24. | Proper submission of all project closeout documents | ___ | ___ | ___ | ___ | ___ |
| 25. | Safety compliances and safety performance | ___ | ___ | ___ | ___ | ___ |
| 26. | EEO & MBE employment compliances | ___ | ___ | ___ | ___ | ___ |
| 27. | Avoidance of claims and claims posturing | ___ | ___ | ___ | ___ | ___ |
| 28. | Avoidance of need for backcharges | ___ | ___ | ___ | ___ | ___ |
| 29. | TOTAL | ___ | ___ | ___ | ___ | ___ |

Describe the most significant good and/or poor performances: ______________________________
______________________________________________________________
______________________________________________________________
______________________________________________________________
______________________________________________________________

Prepared By: ______________________________ Date: ______________

## 4.20
## Sample Project Closeout Checklist

1. All systems on operation
   a. Plumbing ____
   b. HVAC ____
   c. Electrical ____
   d. Fire Protection ____
   e. ________________ ____
   f. ________________ ____

2. Performance tests conducted
   a. Plumbing ____
   b. HVAC ____
   c. Electrical ____
   d. Fire Protection ____
   e. ________________ ____
   f. ________________ ____

3. O & M Manuals delivered
   a. Plumbing ____
   b. HVAC ____
   c. Electrical ____
   d. Fire Protection ____
   e. ________________ ____
   f. ________________ ____

4. Operating Instructions to Owner performed
   a. Plumbing ____
   b. HVAC ____
   c. Electrical ____
   d. Fire Protection ____
   e. ________________ ____
   f. ________________ ____

5. Final completion of physical work
   a. Company Punchlist complete ____
   b. Arch/Owner Punchlist complete ____
   c. All sign-off forms completed ____
   d. Completion Certificate(s) rec'd ____

6. Demobilization complete
   a. Field Offices ____
   b. Equipment & furnishings ____
   c. ________________ ____
   d. ________________ ____

7. Termination of Temporary Services
   a. Hear, light, power, phone ____
   b. Fire, police, guard service ____
   c. Insurance transfers ____
   d. ________________ ____
   e. ________________ ____

8. Final Cleaning of all areas completed
   a. ________________ ____
   b. ________________ ____
   c. ________________ ____
   d. ________________ ____

9. As-Built Drawings
   a. General ____
   b. Site ____
   c. Plumbing ____
   d. HVAC ____
   e. Electrical ____
   f. Fire Protection ____
   g. ________________ ____

10. Guarantees & Warranties
    a. General ____
    b. All subcontractor documents ____
    c. Roof & other bonds ____
    d. ________________ ____

11. Inspection Certificates
    a. ________________ ____
    b. ________________ ____

12. Material/Installation Certificates
    a. ________________ ____
    b. ________________ ____

13. Lien Waivers & General Release
    a. General ____
    b. All subcontractors documents ____
    c. ________________ ____

14. Billing & Charges Processed
    a. All acknowledged Change Orders ____
    b. All subcontractor changes ____
    c. All subcontractor backcharges ____
    d. Final billings from all Sub-Vendors ____
    e. Final billing submitted to Owner ____

15. Steps taken to finalize outstanding claims
    a. To Owner ____
    b. To subcontractors ____
    c. By subcontractors ____
    d. ________________ ____

16. Project Completion Report submitted ____

17. Proj. records transferred to home office ____

18. Forwarding address confirmed
    a. Post Office notified ____
    b. All subs & suppliers notified ____
    c. ________________ ____

19. Other general contract, subcontract, purchase order, specification, and Company Procedure items necessary to close out the project:
    a. Attach list ____
    b. ________________ ____

Section

# 5

# Site Superintendence

## 5.1 The Site Superintendent Function—Section Description

The Superintendent is the key individual responsible for administration of the actual construction in the field. The physical activity on the site will be the source of potentially great successes, or extreme liabilities. The Superintendent, accordingly, must not only be a competent constructor, but must develop and maintain a profound respect for the necessary contributions of each member of the project team. Documentation must be treated as a *fundamental component* of the Superintendent's responsibility, and not as an inconvenient second effort.

Thus the Superintendent must be a competent building constructor *and* administrator as well as possessing significant skill as a team builder. He or she must be able to develop confidence among the direct employees and Subcontractors on the site, mold them all into a cohesive work force, and keep the project always moving forward without actually doing other persons' work for them.

The relationship of the Superintendent function within the organization is outlined in Section 1.5.5 of the Manual. This section details the specific activities that are necessary to actually perform the project Superintendent's responsibilities. Sample full-size forms, checklists, and specific instructions are provided at each step of the way.

This section is very closely coordinated with:

- Section 4, *Project Engineering*
- Section 6, *Safety and Loss Control*
- Section 7, *Progress Schedules and Funds Analysis*

## 5.2 Responsibilities of the Site Superintendent

The Site Superintendent is the individual or team that is directly responsible for the expeditious completion of the physical work. As project engineering in a sense builds the project on paper, the Site Superintendent orchestrates the actual construction. The general duties of the Site Superintendent function, as first outlined in Section 1.5, are developed here as follows:

- Reporting to the Project Manager and carrying out his or her directives with respect to field operations
- Working to assure adequate staffing of the work force, supplies of materials, and complete information as necessary for assembly
- Planning for staffing, materials, and information in advance so as not to interfere with the progress of any one component
- Generating, securing, and otherwise confirming all information needed to create, monitor, and modify the progress schedule on a continuing basis
- Developing the progress schedule with the Project Manager and Project Engineer
- Participating in scope reviews of the various bid packages in order to properly coordinate their respective interfaces and to ensure that nothing is left out and nothing is bought twice
- Working with the Project Manager to develop and administer the site utility plan, site services, security arrangements, safety program, and other facili-

ties and arrangements necessary for appropriate service to the construction effort

- Identifying field construction and work sequence considerations when finalizing bid package purchases
- Monitoring actual versus required performance by all parties; working to bring deviating performances back in line
- Determining whether all Subcontractors are providing sufficient work force and hours of work to actually achieve committed promises of performance
- Monitoring the performance of the Company's purchasing and project engineering efforts to ensure that all subcontracts, material purchases, submittals, deliveries, clarifications, and changes are processed in time to guarantee their arrival at the jobsite by or before the time needed
- Directing any Company field staff
- Being thoroughly familiar with the requirements of the general contract, and thereby identifying changes, conflicts, etc., that are beyond the scope of Company responsibility
- Preparing daily reports, job diaries, narratives, backcharges, notice documentation, and other special documentation as detailed in this and other sections of the Manual, and as may be determined by the Company and by project needs

Although this section includes those specific activities that fall within the restricted definition of the Site Superintendent function, actual practice will necessarily dictate large amounts of overlap of individual responsibilities across the related disciplines noted in Section 5.1. Particular project staffing assignments will determine the actual individuals responsible to *perform* a given function, such as scheduling, or even completing daily field reports. It is important, however, that the Site Superintendent realize that even if it is not his or her direct responsibility to generate the actual document, the Site Superintendent must know what it is, when and how it must be done, who must be doing it, and exactly what his or her contribution to the complete effort must be.

## 5.3 Field Organization

### 5.3.1 Field Staff Considerations

In most instances, projects of moderate size will require a single individual on-site as the Construction Superintendent. Requirements of the function, however, are becoming more demanding every day. Expectations of performance and the actual number of responsibilities are growing well beyond what they were even a few years ago.

Staffing requirements are no longer simply based upon the physical size or relative direct cost of a project. Relatively, the projects themselves may not be becoming any more physically complex, but problems with such factors as

- Quality of design documents
- Subcontractor and Supplier performance

- "Creative" contract provisions
- Claims consciousness of *everyone*
- Exponential increases in every type of liability

now demand especially careful consideration of the complete set of responsibilities that must be properly accommodated in the field every day. Nowhere else but in the position of the Site Superintendent is the principle of "false economy" more evident if the position is skimped on in terms of competence and actual expectations of the individuals who must live up to the role.

The result of these considerations may, in the final analysis, actually turn out again to be the assignment of a single individual to fulfill the characteristic position. In other cases it may be more effective to design the site staffing more closely around other significant project considerations. Accordingly, before the site staff is finalized, consider the following ideas:

1. Site logistics
   *a.* A single building on a small site?
   *b.* A single building on a large, complex site?
   *c.* Multiple buildings at a single location?
   *d.* Multiple buildings at several addresses?
2. Physical size
   *a.* Multiple stories?
   *b.* Size and configuration of building footprint?
3. Site complexities
   *a.* Large open areas (construction ease, material staging, etc.)?
   *b.* Tight site, adjacent buildings?
   *c.* Open or restricted site access for workers and operations?
   *d.* Amount of contact with the public?
4. Project complexity
   *a.* Simple warehouse?
   *b.* Tenant fit-up?
   *c.* Office building?
   *d.* Medical or scientific facility?
   *e.* Low- or high-tech factory or assembly plant?
   *f.* Trade school—complex equipment?
   *g.* "Smart" buildings—complex computer, electrical, and mechanical systems?
5. Building/site relationship
   *a.* Building project with moderate site?
   *b.* Easy location with no problems on adjacent properties?
   *c.* Large, complicated site?
   *d.* Unusual site designs and other needs that "force" the building into the site?
6. Personality and talent of the Superintendent candidate(s)
   *a.* "Pusher" with lower administrative skills?
   *b.* Good administrative talent with weakness in dealing with Subcontractors?
   *c.* Project Manager in Superintendent's clothing?

7. Practical and technical administrative requirements of project
   *a.* Field/home office logistics?
   *b.* Subcontract/direct-hire structure?
   *c.* Level of support expected from home office?
   *d.* Daily and other periodic reporting requirements?
   *e.* Formal and informal meetings and reporting?
   *f.* Simple or complex inspection procedures?
   *g.* Straight or bureaucratic project authorities?
8. Contract type
   *a.* General contract?
   *b.* Design-build?
   *c.* Construction management (advisory)?
   *d.* Construction management with guaranteed maximum price?
9. Apparent quality of construction documents
   *a.* Do they seem to be complete; have all the right parts?
   *b.* Have they been skimped on in terms of absolute amount of information?
   *c.* Do they seem to be properly coordinated?
   *d.* Are they clear or confusing?
   *e.* Were subbids for the major bid packages consistent, or were prices all over the place?
10. Project risk structure
   *a.* Are there liquidated damages? To what extent?
   *b.* Are there heavy or otherwise unique penalties, either described or *implied?*
   *c.* Is there an available bonus for early completion?
   *d.* Is there "heavy" or "sneaky" contract language giving clues to an unfavorable Owner disposition?
11. Value engineering arrangement
   *a.* Are there contract incentives for cost or time reductions? What kind of effort will be necessary?
12. Project purchasing procedure
   *a.* Is purchasing 100 percent home office centralized?
   *b.* Will key purchases be made from the field?
   *c.* Will field purchases remain only for incidental items?

## 5.3.2 Example Field Staff Arrangements

The following examples are listed as possible field staff arrangements that might accommodate the considerations of the previous section. Although they are most common, they are certainly not conclusive. The project itself and the individuals finally selected will determine the best fit. The examples are arranged in order of increasing complexity:

1. Single Superintendent on-site
   *a.* Off-site Project Manager, Project Engineer, Scheduler, and Project Accountant
   *b.* Supplemented with area and/or Assistant Superintendents as necessary

*c.* Moderate administrative skills and disposition
*d.* Strong field tendencies; good with direct work and Subcontractor communications

2. On-site Project Engineer–Superintendent team

*a.* Off-site Project Manager and Project Accountant
*b.* On- or off-site scheduling; preferably on-site
*c.* Allowing the Superintendent to focus totally on field activities with moderate reporting, while major administration is conducted by Project Engineer

3. On-site Project Engineer, Superintendent, and Secretary

*a.* Advantages of example 2, with more complete administrative capability added

4. On-Site Project Manager/Engineer, General Superintendent, and Secretary

*a.* Single individual as Project Manager/Project Engineer to allow project management attention for single project
*b.* Moves more direct authority onto the site, along with complete administrative responsibility
*c.* Scheduling done on-site

5. On-site Project Manager, Project Engineer, General Superintendent, and Secretary

*a.* Intensified arrangement of example 4
*b.* Office Engineers, Area/Assistant Superintendents, and Time Keepers added as needed
*c.* Off-site project accounting authority

6. On-site Project Manager, Project Engineer, General Superintendent, Project Accountant, and Secretary

*a.* Final conversion of example 5 into a complete operational center with profit/loss responsibilities
*b.* Purchasing on- or off-site

## 5.4 Site Utilization Program

### 5.4.1 General

The site utilization program identifies the comprehensive treatment of every use of the site throughout the construction period. Its purpose is to develop, coordinate, and police compliance with all site usages related to the construction, administration, safety, neighborhood relations, and any other physical administrative consideration necessary for the proper logistics of meeting all the practical, legal, and simply considerate measures necessary to work the project.

The complexity, public relations, neighborhood, site, political, and other characteristics of the project may be significant determinants of the approach that will be necessary to confirm the site utilization program that will finally be implemented. In most cases, development of the complete site utilization program will appropriately be left in the hands of the Company project team, with certain courtesies extended to the Owner or major Subcontractors for at least their input. In other cases, major portions of the program may be dictated by the permit, specific Owner concerns, or by the contract documents themselves.

### 5.4.2 Program Components

Components of the complete site utilization program include everything necessary to administer the management, inspection, safety, loss control, security, material staging and handling, site access, and personnel, and all construction support facilities and operations.

Ideally, each facility itself, its location and operation/utilization, will be determined in such a way as to enhance its coordination with the project, and to minimize or eliminate its interference with any part of the project. We do not, for example, want to place the field office in a location that will eventually have to be dug up for a utility line.

Consider each of the items listed here for their immediate need and for your long-term plans. Although each item in the list may appear to be separate and distinct from the others, each is directly affected by every other; so each idea must be considered in complete consideration of all others together. Add to the list as appropriate for a specific project.

1. Configuration of field offices
   - *a.* Quantity and sizes: separate or combined offices for Company, inspectors, design professionals, major Subcontractors
   - *b.* Configurations and composition: office/conference/storage areas, locations of doors and windows, views of site, storage, parking areas
   - *c.* Access to storage rooms
   - *d.* Life safety provision: walkways, entry platforms, handrails, perimeter lighting
   - *e.* Sanitary provision: hookup of trailer facility, use of rental space, separate portable facility
2. Location of field offices
   - *a.* Proximity to construction: close enough, but not too close
   - *b.* Availability of utilities: electric, gas, oil, phone, sanitary; close or need to be brought in
   - *c.* Proximity to adjacent properties: next to the property line, operating businesses, other construction
   - *d.* Proximity to existing parking or temporary construction parking
   - *e.* Proximity to each other: consolidate location of office complex for efficiency in communication, utilities management, etc., or spread out because of narrow site constraints
3. Location of storage trailers; location and configuration of material staging areas
   - *a.* Proximity to construction: close enough, but not too close
   - *b.* Fit within planned security arrangements
   - *c.* Need to be altered at any point during construction
   - *d.* Volume developments and changing needs as construction rates change: starts, increases, and winds down
   - *e.* Ab ility to consolidate (security?), or need/advantage in arranging in different areas (construction ease?)

4. Major stockpile and staging areas
   - *a.* Loam stockpile (left on-site or moved off?)
   - *b.* Structural steel or concrete staging areas
   - *c.* Special materials or equipment
   - *d.* Proximity to hoisting arrangements
5. Site security
   - *a.* Gates and perimeter fencing
   - *b.* Existing, temporary, or new site lighting (which will be incorporated into the project)
   - *c.* Any existing security arrangements
6. Site access
   - *a.* Existing permanent roads and parking
   - *b.* Temporary construction access (into, around, through site): Construction materials and equipment, construction administration, construction personnel
   - *c.* Future permanent access (to be incorporated into project): Roads, parking areas, walks
   - *d.* Time limitations: available hours of day for certain areas, heavy traffic during rush hour, availability on weekends and holidays
7. Safety
   - *a.* Protection of public: covered walkways (with lighting?), barricades, fencing, signs and notices, special lighting, provision for visually or physically handicapped
   - *b.* Protection of Owner or adjacent properties: tree fencing, surface protections
   - *c.* Protection of workers: temporary walkways, barricades, personnel distribution
8. Environmental
   - *a.* Protection of wetland areas: silt fence, buffer zones
   - *b.* Management of hazardous waste
9. Construction waste management
   - *a.* Dumpster locations
   - *b.* Trash chutes
   - *c.* Site-work debris
10. Material hoisting and movement
    - *a.* Truck, tower, or combination cranes: stationary positioning or movement
    - *b.* Material hoists (combined with personnel hoists or separate?)
    - *c.* Changing needs and uses at different stages of construction
11. Restricted areas
    - *a.* Off limits to any construction personnel at any time
    - *b.* Partially restricted: hours of day
12. Interferences with construction
    - *a.* Will any items need to be moved or modified at any point to allow construction to proceed: construction of site underground/overhead utilities (power, storm, sanitary, communications, etc.), parking or roadway areas, walks, playing fields, access to otherwise closed-off areas

### 5.4.3 Sample Site Utilization Plan *(page 5.12)*

*(page 5.12)*

On projects of small to moderate size it may be sufficient to represent components of the site utilization program on the site plan, site utilities plan, or even landscaping plan of the contract set.

Preparing a separate site utilization plan, however, has distinct advantages. Its development will:

1. Aid in the development and coordination of the program components themselves
2. Provide the vehicle to communicate the program to the Owner, design professionals, all Subcontractors, and all authorities having jurisdiction over the work

The Sample Site Utilization Plan that follows is an example of a way to illustrate the properly coordinated program components of the site utilization program. Two points to note in its preparation include:

1. For clarity, it is most often useful to prepare a drawing or sketch of the site and major building components separate from any of the contract drawings that include much extraneous information. This drawing might almost be prepared as a schematic.
2. There is no need to have the program on a full-size drawing. Prepare the document on an 11 by 17-inch sheet of paper. Once it is completed, copies and distribution can be made with a regular photocopy machine; no need to have prints made.

   *a.* The plan can even be reduced to 8½ by 11 inches if such a reduction does not interfere with reading the information.
   *b.* Use of these reduced sizes makes posting and distribution of the program easy and inexpensive.

## 5.5 Field Office Mobilization

### 5.5.1 General

There is much work to be done to establish Company presence in the field. The relative amounts of each step are dependent upon the actual project and staffing considerations as described in Section 5.3, but the specific procedure is essentially the same for each mobilization.

Jobsite mobilization is essentially divided between the actual physical preparation for construction and its administration—what we typically think of when we consider "mobilization." But it includes also preparation of the management of the construction *contract.*

This section, then, details the specifics of:

- Setting up the physical space of the field office
- Arranging for utilities and other site services

**5.4.3**
**Sample Site Utilization Plan**

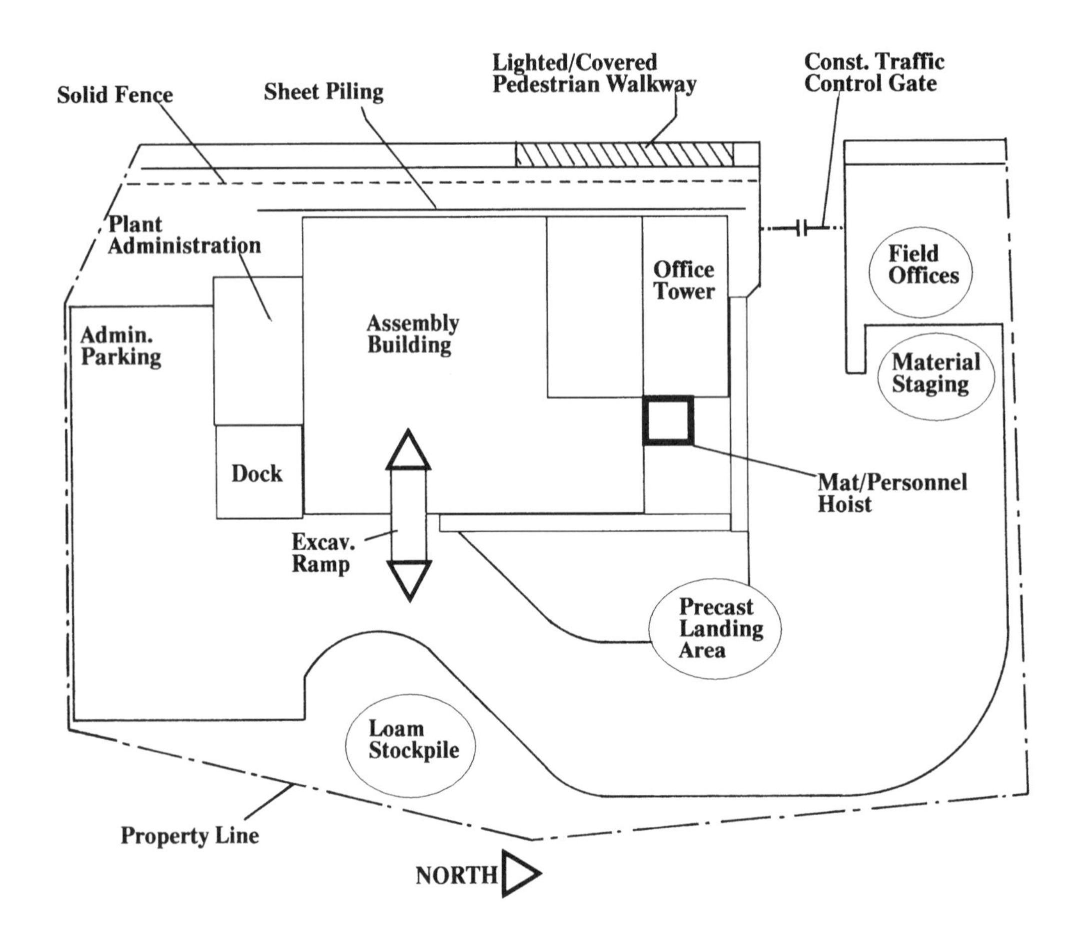


- Establishing contact with Owner and design representatives
- Establishing contact with enforcement authorities

Setting up the administration of the office in preparation for the correct administration of the general contract itself, and of each subcontract package, is dealt with in Section 5.6, *Jobsite Administrative Mobilization.* This section is directly related to and should be closely coordinated with Section 5.4, *Site Utilization Program.* There the relationship of the field office with the entire site is developed.

Finally, this section discusses only the establishment of Company presence, and gives no consideration to any facilities that may be required for Owner, agency, or design professional representatives.

The order in which the process is presented is not important, but all the pieces are. The procedures will be consolidated in Section 5.7, *Jobsite Mobilization Checklist,* with those listed in Section 5.6, *Jobsite Administrative Mobilization,* as a convenient mechanism to help in pushing through the procedure as methodically as possible, while reducing the chance of overlooking any component.

### 5.5.2 Establishing the Field Office and Facilities

Having coordinated the configuration and location of the field office in accordance with Section 5.4, *Site Utilization Program,* proceed with establishing the Company site presence with the following considerations:

1. Field office space

   *a.* Is space available in an existing building (renovation project)? If so, consider problems with construction interference by the office presence.

   *b.* Is retail space available close by?

      (1) Size?
      (2) Cost?
      (3) Proximity to site?
      (4) Level of convenience or inconvenience?
      (5) Expansion or storage possibilities?

   *c.* For items *a* or *b,* will a temporary or conditional Certificate of Occupancy be necessary?

      (1) For construction field staff use?
      (2) If administrative, secretarial, or staff personnel are added to the site?

   *d.* Will office trailer(s) be used?

      (1) Owned or rented, real or assigned cost?
      (2) Size and condition?
      (3) Configuration: office, conference, equipment or supply storage areas?
      (4) Adequate lighting?
      (5) Safe, OSHA-approved stairs, handrails, etc.?
      (6) Adequate heat, power, and air-conditioning?
      (7) Any minimum requirements listed in the specifications?

2. Office furniture

   *a.* Desks, chairs
   *b.* Conference table
   *c.* Folding chairs (for meetings)
   *d.* Plan rack
   *e.* Plan table
   *f.* File cabinets (locking)
   *g.* Book shelves
   *h.* Tackboards (schedule and notice postings)
   *i.* Plan storage cabinets or files
   *j.* Supply cabinets

3. Office equipment and supplies

   *a.* Copier
   *b.* Fax
   *c.* Typewriter
   *d.* Communication equipment
   *e.* Computer and printer (modem?)
   *f.* Blueprint machine or printing arrangements
   *g.* Refrigerator, coffee machine, supplies
   *h.* Bottled water
   *i.* Beepers or pagers
   *j.* Copy, computer, and fax paper

*k.* Maintenance items for all equipment

4. Safety and security equipment

   *a.* Fire and intrusion alarm system
   *b.* Fire extinguishers
   *c.* Hard hats—company personnel and visitors
   *d.* First-aid kit and supplies
   *e.* Emergency phone numbers
   *f.* Stretcher

5. Signs and notices

   *a.* Company "Field Office" signs
   *b.* Visitor "Sign-In" notices
   *c.* Prevailing wage notifications
   *d.* "Hard Hat" signs
   *e.* EEO notifications
   *f.* "Keep Out" and "Restricted Area" notices

6. Office utilities (independent or related to overall site utilities?)

   *a.* Heat—electric, natural gas, propane, oil?
   *b.* Light and power
   *c.* Phone
   *d.* Sanitary

      (1) Use of existing
      (2) Trailer hookup
      (3) Portable units

Once the office space and project administrative areas have been established, start the habit *now* of keeping the entire area *clean and presentable* at all times. This is not just a "good idea"; it is absolutely necessary to:

- Keep personnel work areas efficient
- Preserve office equipment (copiers, faxes, computers, etc.)
- Maintain a safe, healthy environment
- Offer courtesy to others using the space
- Enhance the Company image

### 5.5.3 Establishing Company Presence

During the development of the site utilization plan of Section 5.4 it is a good idea to contact the building officials to at least request the "benefit of their experience" with respect to office location. This is particularly appropriate with the local fire marshal in consideration of life-safety issues.

Once you are at the point of actually establishing the field presence, establish contact with all appropriate individuals with whom you will be living over the project's duration. Let these people know who you are, what we're doing, that we care about doing it right, and that we value their concerns, opinions, and overall relationship. These individuals include:

1. Enforcement authorities

   *a.* General building inspector

*b.* Plumbing, HVAC, and fire protection inspectors
*c.* Electrical inspector
*d.* Fire marshal
*e.* Special inspectors

2. Design professionals

*a.* Architect, office and field representatives
*b.* Plumbing, HVAC, and fire protection engineers
*c.* Electrical engineer
*d.* Structural and civil engineers
*e.* Landscape architect
*f.* Special engineers and consultants

3. Owner representatives

*a.* Field inspectors
*b.* District supervisors
*c.* Administrative offices

### 5.5.4 Visitor Control

Establish visitor control procedures immediately.

1. Have the Visitor Sign-In Forms (sample provided in Section 5.5.5) placed in a clipboard hung on the wall adjacent to the field office door.
2. Post the Visitor Sign-In Notice outside the field office, at every site entrance, and at several appropriate areas throughout the jobsite.
3. *Police the procedure.* It is there to be followed. Follow it.
4. Turn in copies of the sign-in sheet *weekly* to the home office to be placed in the General Project File. It will become an important record not only of who has been on the jobsite and when, but also of Company efforts to enforce appropriate management and safety procedures.

It is a good idea to require that the sign-in forms be returned habitually with any field payroll reporting. In this way, the routine is easy to establish and maintain.

### 5.5.5 Sample Visitor Sign-In Form

The Sample Visitor Sign-In Form that follows is to be used on every jobsite—no exceptions. Place a supply in a clipboard and hang the clipboard adjacent to the field office door.

If you know of any visitor who has not properly signed in (such as a building inspector), always request that the person do so. If that fails, immediately complete the record for him or her.

Return copies of the completed sign-in forms *weekly* to the home office for record.

### 5.5.6 Sample Visitor Sign-In Notice

Reproduce the Visitor Sign-In Notice that follows to include the Company's complete name. Post the notice outside the Company field office, at each site entrance, and at every appropriate area around the jobsite.

## 5.5.5
## Sample Visitor Sign-In Form

**Project:** ______________________________ **Proj. #:** ____________________

| Date | Person | Company / Agency | Hard Hat Own / Issued | Time In | Time Out |
|---|---|---|---|---|---|
| | | | | | |
| | | | | | |
| | | | | | |
| | | | | | |
| | | | | | |
| | | | | | |
| | | | | | |
| | | | | | |
| | | | | | |
| | | | | | |
| | | | | | |
| | | | | | |
| | | | | | |
| | | | | | |
| | | | | | |
| | | | | | |
| | | | | | |
| | | | | | |
| | | | | | |
| | | | | | |
| | | | | | |
| | | | | | |
| | | | | | |
| | | | | | |
| | | | | | |
| | | | | | |
| | | | | | |
| | | | | | |
| | | | | | |
| | | | | | |
| | | | | | |
| | | | | | |
| | | | | | |
| | | | | | |
| | | | | | |
| | | | | | |
| | | | | | |
| | | | | | |
| | | | | | |
| | | | | | |
| | | | | | |
| | | | | | |
| | | | | | |
| | | | | | |
| | | | | | |
| | | | | | |
| | | | | | |
| | | | | | |
| | | | | | |
| | | | | | |
| | | | | | |

**5.5.6**
**Sample Visitor Sign-In Notice**

| ALL VISITORS |
|---|
| Please Report to<br>The (COMPANY) Field Office<br><br>BEFORE Proceeding<br>Anywhere in the Building<br>Or on the Site<br><br>Thank You For Your Cooperation |

## 5.6 Jobsite Administrative Mobilization

### 5.6.1 General

Section 5.5 describes the procedure for what we commonly consider to be "jobsite mobilization," but it is only the establishment of the field presence. Here mobilization is completed by setting up the field office administratively to prepare the field staff for the management of the construction.

The order of the items listed here is not important, but all the components are.

These procedures are consolidated in Section 5.7, *Jobsite Mobilization Checklist,* with those listed in Section 5.5, *Field Office Mobilization,* as a more convenient mechanism to help push through the procedure as methodically as possible.

### 5.6.2 Construction Administration

This section lists the supply of job forms to be assembled at the job start-up, which will be necessary to routinely complete all records and reporting as required by the respective sections of this Manual, and provide for the establishment of the project filing system. Each item listed here and its use are described in detail in the appropriate sections of the Manual:

1. *Project manuals and log books*
   *a.* Company Operations Manual

*b.* Any specific project operations manual
*c.* Subcontractor Summary and Telephone Log
*d.* Submittal Log (if submittals processed on-site)
*e.* Change Order Summary Log (if change order administration is done on site)
*f.* RFI Log

2. *Supply of administrative forms.* Most, if not all, of the forms listed here are provided in the Company Operations Manual. This list is produced as a convenience to ensure that an appropriate supply of each form is provided at job start-up.

*a.* Daily Field Reports
*b.* Visitor Sign-In Forms
*c.* Change Order Forms
*d.* Quotation and Phone Quotation Forms
*e.* Field and Administrative Payroll Forms
*f.* Time and Material Tickets
*g.* Job Meeting Forms
*h.* Record of Meeting/Conversation Forms
*i.* Memos
*j.* Photograph Record Forms
*k.* Excavation Notification Checklists
*l.* Equipment Use Release Forms
*m.* Full and Partial Lien Waiver Forms
*n.* Transmittal Forms
*o.* Fax Memo/Transmittal Forms
*p.* Field Purchase Order Forms
*q.* Request for Information (RFI) Forms
*r.* Certified Payroll Report Forms
*s.* Schedule Status Report Forms
*t.* Backcharge Forms
*u.* Subtrade Performance Evaluation Forms

3. *Start-up project files.* Set up the files as described in Section 2.4, *Files and File Management.* Perform the filing procedure strictly in accordance with the requirements of that section. The Project Manager or his or her designee at the home office is to be copied on *all* correspondence generated from the field. File designations for the respective item at the home office will then be determined by the person there.

Although maintenance of the field files is a fundamental responsibility of the field staff, file management must proceed with the idea that the home office files *must* be maintained in a current and absolutely complete manner at all times.

The jobsite files are to be kept in *locking* file cabinets. After the specific folders have been set up, start-up files will include:

*a.* Contract and correspondence files
*b.* Subcontracts and Purchase Orders for all bid packages
*c.* Certificates of insurance for *everyone* allowed to work on the site
*d.* Approved submittals for any construction about to proceed
*e.* Executed Equipment Use Release Forms as appropriate
*f.* Any relevant correspondence in existence at start-up

4. *Construction and contract documents.* Arrange for copies of all documents relevant to every aspect of the technical construction, including:

*a.* The plans, specifications, and all addenda—Construction Set

*b.* The plans, specifications, and all addenda—As-Built Set
*c.* Project Manual or working procedure instruction
*d.* Current copies of relevant building codes
*e.* Copies of referenced standard specifications
*f.* All other special project documentation as required

5. *Contract management documents.* Arrange for copies and be aware of the specific details of all contractual responsibilities of the Company—in addition or as a supplement to the technical construction responsibilities:

   *a.* The Owner or Company agreement
   *b.* Bid documents

6. *Overall contract considerations.* Be aware of the general complexion of the Owner or Company agreement. Know the contract type and overall parameters of expected performance, including:

   *a.* Contract type (General Contractor, Design-Build, CM, CM w/GMP, etc.)
   *b.* Contract execution date
   *c.* Construction start date
   *d.* Completion date(s)
   *e.* Number of calendar days
   *f.* Number of working days
   *g.* Presence of liquidated or other "damage" provisions; if so, what?
   *h.* Presence of bonus provisions
   *i.* Unusual restrictions (physical or contractual)

7. *Contract execution.* Be completely familiar with all project start-up requirements and procedures, and those to be routinely managed throughout the project, including:

   *a.* General Building Permit—who secures, who pays?
   *b.* Plumbing, HVAC, fire protection, electrical, and other permits—who secures, who pays?
   *c.* Billing procedure:

      (1) Date Subcontractor requisitions due
      (2) Date general requisition due to Owner
      (3) Status of the General Schedule of Values
      (4) Status of Sub-Vendor Schedules of Values
      (5) Requisition review/approval procedure

   *d.* Change order procedure

      (1) Change clause present
      (2) Procedure and forms required

   *e.* EEO requirements

      (1) Mandatory
      (2) Good faith

   *f.* Independent testing laboratories

      (1) Areas and work types required
      (2) Payment responsibilities

   *g.* Baselines and benchmark

      (1) Responsibility to provide
      (2) Responsibility for engineering expense

*h.* Job meeting arrangement and schedule

(1) Who provides the minutes

*i.* Dispute resolution provisions

(1) Dispute resolution clause or prescribed procedure present
(2) Arbitration provision
(3) Notice requirements

*j.* Notice requirements for various circumstances

8. *Job cost and production control.* Be aware of all project cost considerations and your complete responsibilities relative to the achievement of all cost/progress targets. Relevant documents include:

*a.* General project budget
*b.* Resource estimates, labor and materials
*c.* Job cost report
*d.* Change order estimates
*e.* Baseline construction schedule
*f.* Baseline cash-flow projection

9. *Documentation of the site utilization program.* The development of the site utilization program and of the plan itself is treated in Section 5.4. Here the program is to be posted and distributed, and arrangements are to be made to secure the relevant provisions of the plan. These are site considerations, as distinguished from the field office considerations of Section 5.5.

*a.* Temporary fences and other protective and security measures
*b.* Guard service, security lighting, security systems, or other appropriate security determinations
*c.* Temporary electric
*d.* Temporary water
*e.* Dumpster, waste-disposal arrangements
*f.* Arrangements for management of hazardous materials
*g.* Progress photograph arrangements
*h.* Testing laboratory service arrangements

(1) Soils
(2) Concrete
(3) Steel and welding
(4) Other/special

## 5.7 Jobsite Mobilization Checklist

The checklist that follows summarizes the discussions and outlines of Section 5.4, *Site Utilization Program*; Section 5.5, *Field Office Mobilization*; and Section 5.6, *Jobsite Administrative Mobilization.* It is provided to facilitate following through on the arrangements for the many decisions, services, suppliers, and facilities described throughout the complete mobilization process.

Some of the forms listed are described in later sections of the Manual. For these and all others, refer to the individual sections for more complete discussions of the respective items and their relevance to all others. Make a copy of the checklist and use it throughout the start-up of every project.

# 5.7
# Jobsite Mobilization Checklist
# (Page 1 of 3)

SITE & SITE SERVICES

1. Site Utilization Plan Prepared ____
   Approved by: ____________ ____
2. Temp. fences, protection (see safety) ____
3. Guard service ____
4. Temporary electric ____
5. Temporary water ____
6. Dumpster, disposal arrangements ____
   Who pays: ____________ ____
7. Progress Photograph service ____
   Who pays: ____________ ____
8. Testing Laboratories
   a. Soils____
   b. Concrete ____
   c. Steel/welding ____
   d. Other: ____________ ____
   Who pays: ____________ ____
9. Weather information phone numbers ____
10. Call-Before-You-Dig" or other One-Call System in effect: ____________ ____

FIELD OFFICES & OFFICE EQUIPMENT

1. Facility type
   a. Trailers ____
   b. Retail space ____
   c. Space within the construction area ____
   d. Other: ____________ ____
2. Number of Offices
   a. Company ____
      1) Project Manager ____
      2) Superintendent ____
      3) Other: ____________ ____
   b. Owner representative(s) ____
   c. Design professionals ____
   d. Subcontractors ____
      1) ____________ ____
      2) ____________ ____
      3) ____________ ____
      4) ____________ ____
      5) ____________ ____
      6) ____________ ____
3. Temporary Facilities
   a. Heat (Type:____________) ____
   b. Lighting & Power ____
   c. Telephones/portable phones ____
   d. Site communications equipment ____
   e. Lavatories
      1) Use of existing ____
      2) Hookup of trailers ____
      3) Portable (Quantity: ______) ____
   f. Water
      1) Use of existing ____
      2) Hookup of trailers ____
      3) Bottled Water ____
   g. Other:____________ ____
   ____________ ____
4. Office Furniture
   a. Desks, chairs, stools ____
   b. Conference table ____
   c. Folding chairs ____
   d. Plan rack ____
   e. Plan table ____
   f. File cabinets
      1) Regular ____
      2) Locking ____
      3) Fireproof ____
   g. Bookshelves ____
   h. Plan storage cabinets ____
   i. Supply cabinets ____
   j. Plan edge reinforcing machine ____
5. Office Equipment & Supplies
   a. Copier ____
   b. Fax ____
   c. Typewriter ____
   d. Communication equipment ____
   e. Computer, Printer & Modem ____
   f. Software
      1) Word Processing ____
      2) Scheduling ____
      3) Spreadsheet ____
      4) Database ____
      5) Other: ____________ ____
   g. Printing arrangements ____
      1) Blueprint machine ____
      2) Printing arrangements ____
   h. Refrig., coffee mach., supplies ____
   i. Beepers, pagers ____
   j. Copy, computer, fax paper ____
   k. Maintenance items for all equipment ____
6. Office Safety & Security Equipment
   a. Fire & Intrusion Alarm ____
   b. Fire extinguishers ____
   c. Hard hats
      1) Company personnel ____
      2) Visitors ____
   d. First Aid Kit & supplies ____
   e. Emergency phone numbers ____
   f. Stretcher ____
   g. Names of any employees with medical training
      1) ____________
      2) ____________
      3) ____________
7. Signs & Notices
   a. Company "Field Office" signs ____
   b. Visitor "Sign-In" Notices ____
   c. Prevailing Wage or other Labor Department Notifications ____
   d. "Hard Hat" signs ____
   e. EEO Notices ____
   f. "Keep Out," "Danger," and "Restricted Area" Notices ____

# 5.7
# Jobsite Mobilization Checklist
# (Page 2 of 3)

ADMINISTRATION

1. Project Manuals & Log Books
   a. Company Operations Manual ____
   b. Project Operations Manual ____
   c. Sub. Summary & Phone Log ____
   d. Submittal Log ____
   e. Change Order Summary Log ____
   f. RFI Log ____
2. Supply of Job Forms
   a. Daily Field Reports ____
   b. Visitor Sign-In Sheets & Clipboard ____
   c. Change Order Forms ____
   d. Quotation & Phone Quote Forms ____
   e. Field Payroll Forms ____
   f. Administrative Payroll Forms ____
   g. Time & Material Tickets ____
   h. Job Meeting Forms ____
   i. Record of Mtg/Conversation Forms ____
   j. Memos ____
   k. Photograph Record Forms ____
   l. Excavation Notification Checklists ____
   m. Equipment Use Release Forms ____
   n. Full & Partial Lien Waiver Forms ____
   o. Certified Payroll Report Forms ____
   p. Transmittal Forms ____
   q. Fax Memo/Transmittal Forms ____
   r. Field Purchase Order Forms ____
   s. Request for Information (RFI) Forms ____
   t. Backcharge Notices ____
   u. Backcharge Forms ____
   v. Schedule Status Report Forms ____
   w. Sub Performance Evaluation Forms ____
3. Start-Up Project Files
   a. Contract & Correspondence Files ____
   b. Submittal Files ____
   c. Special Files ____
4. Start-Up Subcontractor Submissions
   a. Subcontracts & Purchase Orders ____
   b. Certificates of Insurance ____
   c. Sub Payment & Perform. Bonds ____
   d. Executed Equip. Use Release Forms ____
   e. Approved Submittals ____
   f. Other: ____________ ____
5. Project Directory ____

CONSTRUCTION & CONTRACT DOCUMENTS

1. Plans, Specs, Addenda - Const. Set ____
2. Plans, Specs, Addenda - As-Built Set ____
3. Project Manual/Working Procedure ____
4. Building Codes ____
5. Referenced Standard Specifications ____
6. Other:
   1) ____________ ____
   2) ____________ ____
   3) ____________ ____

CONTRACT MANAGEMENT DOCUMENTS & GENERAL INFORMATION

1. Owner/Company Agreement ____
2. Bid Documents ____
3. Contract Type (GC, CM, CMw/GMP, DB) ____
4. Contract Execution Date: ____________
5. Construction Start Date: ____________
6. Substantial Compl. Date: ____________
7. Final Completion Date: ____________
8. Number of Calendar Days: ____________
9. Number of Working Days: ____________
10. Liquidated Damages ____
    a. Value/Day: $ ________
11. Bonus ____
    a. Value/Day: $ ________
12. Special Considerations:
    ________________________
    ________________________

CONTRACT EXECUTION

1. Permits — Who Pays — Rec'd
   a. General Building ________ ____
   b. Plumbing ________ ____
   c. HVAC ________ ____
   d. Fire Protection ________ ____
   e. Electrical ________ ____
   f. ________ ________ ____
   g. ________ ________ ____
2. Billing Procedure
   a. Date subcontractor requisitions due ____
   b. Date General requisitions due ____
   c. Requisition review/appr. procedure ____
3. Change Order Procedure
   a. Change Clause: Sect. ________ ____
   b. Procedure/forms required ____
4. EEO Requirements
   a. Mandatory ____
   b. Good Faith ____
5. Independent Testing Laboratories
   a. Areas & work types required:
      1) Soils ____
      2) Concrete ____
      3) Steel ____
      4) Other: ________ ____
   b. Payment responsibility:
      ________________
6. Baselines & Benchmark
   a. Responsibility to provide:
      ________________
   b. Payment Responsibility:
      ________________
7. Job Meetings
   a. Preconstruction Meeting Date: ____
   b. Regular Meeting Schedule ____
   c. Who provides official minutes: ____

## 5.7 Jobsite Mobilization Checklist (Page 3 of 3)

8. Dispute Resolution
   a. Dispute resolution clause:
      Gen. Cond. Section ____________
      Suppl. Cond. Section ____________
   b. Arbitration Provision:
      Gen. Cond. Section ____________
      Suppl. Cond. Section ____________
   c. Notice Period: ____________
   d. Special considerations:
      ____________
      ____________
      ____________

JOB COST & PRODUCTION CONTROL
1. General project budget ____
2. Resource estimates: material & labor ____
3. Job Cost Report ____
4. Change Order estimates ____
5. Baseline Construction Schedule ____
6. Baseline Cash-Flow Projection ____

PROJECT CONTACTS
1. Enforcement Authorities
   a. General Building Inspector:
      ____________
   b. Plbg./HVAC/Fire Protect. Inspector:
      ____________
   c. Electrical Inspector:
      ____________
   d. Fire Marshal:
      ____________
   e. Special Inspectors:
      ____________
      ____________
      ____________
      ____________
2. Design Professionals
   a. Architect: Office Representative:
      ____________
   b. Architect: Filed Representative:
      ____________
   c. Plumbing & HVAC Engineer(s):
      ____________
      ____________
   d. Fire Protection Engineer:
      ____________
   e. Electrical Engineer:
      ____________
   f. Structural Engineer:
      ____________
   g. Civil Engineer:
      ____________
   h. Landscape Architect:
      ____________
   i. Special Engineers & Consultants:
      ____________
      ____________
      ____________
      ____________
3. Owner Representatives
   a. Field Inspector:
      ____________
   b. District Supervisor:
      ____________
   c. Other:
      ____________
      ____________
      ____________
4. Security / Life Safety
   a. Police: ____________
   b. Fire: ____________
   c. Hospital: ____________
   d. Emergency: ____________
   e. Alarm Service: ____________
5. Jobsite Personnel home phone numbers ____

# 5.8 Subcontractor Summary and Jobsite Phone Log Book

## 5.8.1 General

Section 2.4.7 describes the basic preparation of the Jobsite Subcontractor Summary and Phone Log and its coordination with the other project documentation mechanisms. This section details its use.

The binder will be the location in which the summaries of each Subcontractor will be noted and performance compliances with respect to their jobsite responsibilities (as opposed to their contract and engineering responsibilities of Section 4.4) will be consolidated. It will provide at a glance the status of a particular vendor's performance and, more importantly, its nonperformance.

If the scheduling documentation is being correctly maintained, and the Daily Field Reports are being kept as current and as comprehensive as they should be, it may not become necessary to use the phone log feature of this book fully, as it would become a redundant effort. If, however, these kinds of notes are not being added to the Daily Field Reports routinely and consistently, the notes feature of the Log Book *must* be used. The procedure is as follows:

1. Prepare an appropriately labeled 1-inch three-ring binder in a color to match the project set. Insert a set of alphabetically tabbed dividers.
2. Prepare a Jobsite Subcontractor Summary Form, as described hereafter, for each Vendor and place it in the appropriate alphabetical location.
3. As completed Subcontractor/Supplier Reference Forms (see Section 4.5) are received for each Vendor, insert them behind the respective Summary Forms.

   Note that the absence of a form by the time any Vendor is on the site is your cue that either the home office has for some reason neglected to forward you your copy, or the Vendor has not yet completed it and sent it in. In either case, whenever you notice a form to be missing, this is the time to get it.
4. Place a small supply of Telephone Log Forms as described hereafter behind the Summary and Reference Forms for each Vendor.
5. Follow through with the use of each form as described throughout the remainder of this section.

## 5.8.2 Purpose

Each project will have dozens of Vendors, and each Vendor will have dozens of jobsite requirements. A precious few Subcontractors and Suppliers will take the appropriate initiative and shoulder their responsibilities in a well-coordinated, responsible manner. Most, however, will fit somewhere into a wide range of capabilities, tendencies, and intentions.

These Vendors are the ones who will need constant pushing, pulling, and chronic follow-up to ensure not only that they do everything they're supposed to, but also that things are done when and in the manner they're needed by *you.*

The first purpose, then, of the Summary Forms is to consolidate the routine items that must be provided or complied with by each Sub-Vendor. Summarized in this way in a checklist fashion, it becomes very convenient to review performance and compliance, and to identify those cracks through which something might have fallen.

The second purpose is to increase the ease with which the chronology of communication with the various Sub-Vendors can be reviewed and used. Daily Field Reports (see Section 5.9) are the traditional means to document not only on-site performance, but also nonperformances, your expediting efforts, notices given, conversations, and so on. If the daily reports are maintained as instructed in that section, documentation will at least be adequate.

The difficulty with providing records of conversations, notices, and the like on the daily reports is that:

1. The information is consolidated with all other project information for that particular day. It may be buried deep and be difficult to locate.
2. All relevant information for a specific Vendor is spread out over possibly hundreds of pages over the life of the project. It will be there to support your contentions in a claim or quasi-claim situation, but it will never be researched effectively (or at least very conveniently) for practical purposes so that it can be used during the regular daily communication cycles.

Providing this kind of information in the Log Book solves these problems. The Daily Field Reports will speak for the actual field performance. The Log Book will catalog the communication history and consolidate it in a way that will provide at a glance:

1. Everything that needs to be complied with at the moment or in the near future
2. Previous requirements that have not yet been properly met
3. The short- and long-term histories of easy or tedious and strained communications

### 5.8.3 Procedure and Use

1. Use the standard form. Add any specific requirements unique to the project and photocopy a supply of *project* forms.
2. Add any requirements specific to a particular Subcontractor or Supplier to that Vendor's forms. Try to do this as the form is initially being prepared, and develop the habit of considering this idea every time you look at any of the forms.
3. *The most important idea of all is to develop the habit of always having the binder open and in front of you.* This aspect of the Log Book's use is crucial to the realization of its full potential as a management tool and as a documentation vehicle.
4. Use the binder as your regular telephone and address directory. In this way you'll automatically have it open to that Vendor's summary *before you pick up the phone.*
5. Log all relevant information *as it occurs.* Since the binder is already open, it will take less than a moment. Waiting for the end of the week, or even later in the day, is your guarantee that the information will never be recorded.
6. Enter *all* information, especially when nothing has happened. If, for example, all you get for your efforts is "he's not in; he'll call you back," note those remarks along with who gave them, date, and time. A list of six of these in a row, for example, goes a long way in demonstrating the performance failures of the other party.
7. Use every opportunity to "subtly" make everyone aware of these records.

   *a.* *Having* the records is the first step toward getting the upper hand in every situation every time.

   *b.* If the other parties are *aware* of the existence of such detailed records, they're more likely to learn fast that it is in their best interest to avoid games. When they see your ability to simplify greatly the usually extreme complexity of project relationships, they'll take another look. They'll hopefully decide that it's best not to fool with *this* relationship.

Some ways to do this include:

*a.* Whenever a Sub-Vendor phones to complain, to question, or for any other reason:

(1) Review the Vendor's performance summary *before* you pick up the phone.

(2) Before the party gets a chance to say what his or her problem is, fire off *every* piece of paper that has not been delivered properly, and every item that you're still waiting for. Then move into any specific jobsite performance issue that *you* have. Be sure to advise him or her of all those dates (and times) that you've been trying to reach him or her about *your* issues. After you have gone through all *your* issues and documented the latest round of performance commitments relative to them, *then* you can ask what *their* problem was.

*b.* In any formal or informal meeting:

(1) Open the binder whenever a relevant issue is raised.

(2) Let everyone *see* you making your notes in their file *as the discussion is occurring.* This action gives profound legitimacy at a later date to *your* information as being the *right* information. It will make everyone think before they speak, and force them to consider their actions (and inactions) very seriously.

8. If more elaborate confirmation or an impromptu meeting has taken place, consider in addition to your notes using the *Meeting/Conversation Record Form* of Section 4.13.12.

### 5.8.4 Sample Jobsite Sub-Vendor Summary Form—Completed Example *(page 5.27)*

The Sample Jobsite Sub-Vendor Summary Form included here is the form to be used to follow through on the procedure described in this section. The completed form is provided as an example of its proper use. Section 5.8.5 provides the blank form to be photocopied for actual use.

### 5.8.5 Sample Jobsite Sub-Vendor Summary Form—Blank Form *(Page 5.28)*

### 5.8.6 Sample Telephone Log Form—Completed Example *(page 5.29)*

The Sample Telephone Log Form included here is the same form that was used in Section 4.4 by the Project Engineer. It is to be used to follow through on the procedure described in this section. A completed form is provided as an example of it proper use. Section 5.8.7 provides the blank form to be photocopied for actual use.

### 5.8.7 Sample Telephone Log Form—Blank Form *(page 5.30)*

# 5.8.4
# Sample Jobsite Sub-Vendor Summary (Completed Example)

PROJECT: TRADESMAN SCHOOL

Sub: FOUNDATIONS, INC.
219 AMITE BLVD.
COLWELL, CT 06902

Sect.: 03200
03300

Company #: 9890

Owner #: BI-Q-177(B)

Sub-Vendor #: —

Name: JERRY GRANT

Phone: 555-1625

Fax: 555-1670

1. Subcontractor Approval Received Date: 6.4.94
2. Subcontract Executed / On-Site Date: 6.10.94
3. Insurance Certificate On-Site Date: 6.10.94
4. Information Request Form On-Site Date: 6.10.94
5. Schedule of Values Approved / On-Site Date: 6.24.94
6. Plans & Specs Delivered Date: 5.15.94
7. Submittal Requirements Letter Sent Date: 5.15.94
8. Shop Drawings Received: ______ Approved: ______
9. Construction Schedule Delivered: 5.15.94 Confirmed: 6.10.94

10. Required Material Delivery Date Date: 6.14.94
11. Material Delivery Confirmation Date Date: 6.14.94

Correspondent: J. GRANT

Qualifications: SITE START CONFIRMED. FOOTING, FNDTN WALL & S.O.G. START DATES CONFIRMED; ALL ALLOWED DURATION ACCEPTABLE.

12. Material Received:

___ Ahead of Schedule Date: ______

X On Schedule Date: 6.14.94

___ Behind Schedule Date: ______

Remarks:

CONC. DELIVERIES PROGRESSED AS-NEEDED

13. As-Built Drawings Prepared / Updated Date: 6.24.94
Date: 7.12.94
Date: 7.19.94
Date: ______

14. Equipment Use Release Forms Executed Date: 6.13.94

15. Backcharges:

| Date: | Amount: | Description / Remarks: |
|---|---|---|
| 8.10.94 | $1,422.- | REPAIR CRACKED WALL; DAMAGED BY CONC. TRUCK. |
| | | |
| | | |
| | | |
| | | |
| | | |

16. Notes & Remarks (Use Additional Pages as Required):

QUALITY OF WORK & SCHEDULE PERFORMANCE ACCEPTABLE. COMPETENT SUPERVISION & WILLINGNESS TO COOPERATE.

NEEDED HELP CONFIRMING LAYOUT OF THEIR WORK.

GENERALLY ON-TIME W/ PAPERWORK

## 5.8.5
## Sample Jobsite Sub-Vendor Summary

PROJECT: ______________ Sub: ______________ Sect.: ______________

Company #: ______________
Owner #: ______________ Phone: ______________
Sub-Vendor #: ______________ Name: ______________ Fax: ______________

1. Subcontractor Approval Received Date: ______________
2. Subcontract Executed / On-Site Date: ______________
3. Insurance Certificate On-Site Date: ______________
4. Information Request Form On-Site Date: ______________
5. Schedule of Values Approved / On-Site Date: ______________
6. Plans & Specs Delivered Date: ______________
7. Submittal Requirements Letter Sent Date: ______________
8. Shop Drawings Received: ______________ Approved: ______________
9. Construction Schedule Delivered: ______________ Confirmed: ______________

10. Required Material Delivery Date Date: ______________
11. Material Delivery Confirmation Date Date: ______________
    Correspondent: ______________
    Qualifications: ______________

12. Material Received:
    ____ Ahead of Schedule Date: ______________
    ____ On Schedule Date: ______________
    ____ Behind Schedule Date: ______________
    Remarks: ______________

13. As-Built Drawings Prepared / Updated Date: ______________
    Date: ______________
    Date: ______________
    Date: ______________

14. Equipment Use Release Forms Executed Date: ______________

15. Backcharges:

| Date: | Amount: | Description / Remarks: |
|---|---|---|
| | | |
| | | |
| | | |
| | | |
| | | |
| | | |

16. Notes & Remarks (Use Additional Pages as Required):

# Telephone Log Form
# (Completed Example)

**Project:** WESTSTREET SCHOOL RENOVATION **Proj. #:** 96200

| Date | Time | Discussion |
|---|---|---|
| 5·19·94 | 8:30 AM | CALLED FOR RAY DEERING TO ASK WHEN WE WILL RECEIVE |
| | | SUBMITTALS - NOT IN; WILL CALL BACK |
| | 2:20 PM | " - " " " " " |
| 5·20·94 | 9:10 AM | " - RAY IS EXPECTED BY NOON |
| | 1:05 PM | " - RAY CONFIRMED THAT ALL SUBMITTALS WILL |
| | | BE IN THIS OFFICE BY 5·26·94 |
| 5·25·94 | 11:25 AM | CALLED RAY DEERING TO CONFIRM 5/26 DEL. OF SUBMITTALS |
| | | -RAY SAID HE DOESN'T EXPECT FROM HIS SUPPLIER |
| | | "FOR ANOTHER WEEK" |
| | | · I ADVISED RAY THAT EVEN THE 5·26·94 DATE WAS |
| | | A WEEK AFTER HIS SUBCONTRACT COMMITMENT, & |
| | | THIS ADD'L SLIP IS TOTALLY UNACCEPTABLE - ITS |
| | | PUSHING ALL CONCRETE FOOTING WORK BACK BY AS |
| | | MANY DAYS. |
| | | ·RAY SAID "HE'LL SEE WHAT HE CAN DO, & GET BACK" |
| | | |
| | | (REFER TO COMPANY MEMO TO RAY THIS DATE.) |
| 5·26·94 | 9:15 AM | CALLED FOR RAY - NOT IN; EXPECTED BY 2:00 P.M. |
| | 2:30 PM | " " " " " |
| | 4:00 PM | " " " " " |
| 5·27·94 | 8:30 AM | " " " " " |
| | | -ASKED FOR JOHN SUTUR (RAY'S EXEC.) |
| | | -EXPLAINED THE ENTIRE SITUATION & DELAY TO JOHN |
| | | -JOHN SAID "HE'LL GET TO THE BOTTOM OF IT & |
| | | GET BACK TO ME ASAP - HOPEFULLY TODAY" |
| | 11:05 AM | · J. SUTUR CALLED BACK; ADVISED THAT THE SUPPLIER WILL |
| | | EXPRESS MAIL DIRECT - WE SHOULD HAVE IN OUR |
| | | OFFICE BY 10:30 AM TOMORROW |
| | | - I ASKED FOR SUPPLIER'S PHONE # TO VERIFY |
| | | (802) 555-1629 FRED DAVIS. |
| | 11:15 AM | · CALLED FRED, WHO CONFIRMED JOHN'S REMARKS. |
| | | |
| | | (REFE TO CO. MEMO THIS DAT TO BOTH |
| | | FRED DAVIS & JOHN SUTUR) |

# 5.8.7
# Telephone Log Form

**Project:**________________________________________ **Proj. #:** ____________________

| Date | Time | Discussion |
|---|---|---|
| | | |

## 5.9 Daily Field Report

### 5.9.1 Description and Responsibility

Maintaining complete, comprehensive, detailed records of every facet of performance at the jobsite is much more than just a good idea. It is fundamentally essential to the efficient control of the work, to the achievement of all Company and project objectives, and to the control of the potentially extreme liabilities lurking in every corner of the site—including the Owner, design professionals, and all Subcontractors.

The Daily Field Report is the most important project documentation mechanism at the jobsite used to accomplish these objectives. It is to be filled out *every day.* On most projects the report will be completed by the Project Superintendent. On very large projects, the report may be completed by the Project Manager, with field information supplemented by the Superintendent. On very small projects, the report should be maintained by whoever has absolute project responsibility.

### 5.9.2 Purpose

The Daily Field Reports will be the detailed record of precisely what did—or did not—happen on a given day or throughout a particular period. Such detailed history is prepared in such a manner that it may be consulted to confirm the particulars of the facts whenever any portion of the work is questioned, or the performance of any party becomes an issue. It will become the key basis for support in the prosecution or defense of claims in both directions involving the Owner, design professionals, and Subcontractors.

The purposes of the Daily Field Report are therefore to:

1. Provide a chronological, day-to-day account of the work force, the respective activities planned, and the actual activities performed
2. Record all visitors to the site, with any significant comments
3. Provide easily retrievable weather information, pertinent data that may support activity or inactivity
4. Isolate by description specific work to be charged to and otherwise identified with any change order
5. Record materials and equipment being received on or sent from the site by the Company or by any Subcontractor
6. Call attention to particular problems or situations
7. Provide a written request for materials, information, or other assistance from the Company home office or senior management

### 5.9.3 Need for Proper, Consistent Attention

It is important to recognize from the onset that the greatest problem with Daily Field Reports is simply that the information in them is not needed until there is a problem—but then it is needed *badly.* Characteristics that contribute to the problem include the following:

1. Because the reports are not consulted every day (or even periodically if the project has not been plagued with problems), the process can too easily get looked upon by those who prepare the report as one about which the home office or senior management is really not all that concerned.
2. When busy activities around the site begin to steal precious time from the Superintendent, it will become tempting to begin to simplify the information on the reports. If allowed, the process will continue to degenerate:

   *a.* First, they won't really be prepared daily. They'll be "caught up with" near the end of the week, or at some other later period.
   *b.* The accuracy of the information will become compromised. Because of time lag, the edge will be off (was the concrete placed along column line K or J? Was it placed on Monday or Tuesday?).
   *c.* The notes themselves will become short and marginally useful. When a large amount of information needs to be caught up on, it will all be shortened and compromised.

To combat these common traps, the Superintendent and all other individuals even marginally associated with the effort must begin with the profound realization that this reporting responsibility is a *fundamental* requirement of the position. If it is not being done correctly, it is to be considered a *major* performance deficiency.

From that point we must all realize the critical nature of the information. We can accept the idea that there is a high probability that most of the information will never be consulted (if there are no problems), but we must face the reality that when there is a problem, the information that will be needed will be needed in great detail.

Since there is, of course, no way to determine in advance what pieces of information will be needed in such an event, it is crucial to maintain *all* information to acceptable degrees of detail. The potential problem and the critical need for proper perspective and attention by the entire project staff cannot be overemphasized.

## 5.9.4 General Procedure

1. The Project Superintendent must fill out the report every day. Keep the day's report on a clip board, and carry it with you wherever you go. In this way you will have it handy to jot down all relevant information *as it is observed.* As this habit becomes well developed, the daily completion of the report will become almost automatic; by the end of the day it will already have been done.
2. Each report is to be signed by the person initiating the information. If the Project Manager is an on-site position, he or she must also sign the report.
3. Copies of the reports *must* be distributed on a current basis to the home office for periodic review by senior management, and to be filed in the General Project File. Two optional ways to effectively police this procedure are to:

   *a.* Attach copies of the reports to the jobsite Payroll Report Forms being returned to the office each week. Instruct those responsible for processing the payroll to contact the site immediately whenever the Daily Field Reports are not attached, and make every effort to enforce the policy.
   *b.* Require the Superintendent to mail the copy of the Daily Field Report to the home office every day. This really does not add much expense, but has the significant benefit of ensuring that the reports are in fact being prepared daily. This is the preferred procedure.

In either case it is equally the responsibility of the Project Manager or other appropriate senior management to develop their own habit of routinely reviewing these reports. It not only is a great way to become currently aware of the project details, but it will help you to assist the field staff in improving their information recording methods, and in calling attention to missed or otherwise incompleted reports.

### 5.9.5 Report Preparation Guidelines

The following general ideas are to be used throughout the preparation of each Daily Field Report.

1. Be aware of the *Rules of Effective Project Correspondence* as detailed in Section 2.3.2.
2. Report all facts. Keep the report with you at all times, and record *all* relevant information. If there is not enough space in any category, use as many continuation sheets as necessary. Never abbreviate notes because of lack of space—get more space.
3. Report only facts. Keep frustrations, inuendo, and any remarks that do not belong in a professional communication out of it altogether.
4. Draw only appropriate conclusions. Do not draw speculative conjecture. Include only those conclusions that are the result of a direct cause-and-effect relationship, and that lead to or require some action (correction of work, backcharge, etc.).
5. Avoid buzzwords and trade jargon. Use ordinary language to describe everything.
6. Brief is acceptable, but be complete. Outline statements are fine if they still manage to include all facts and complete descriptions.
7. *Be precise.* Note specific locations, limits of work, and whatever information is necessary to identify the work and processes described positively without question. "Poured concrete," for example, does no one any good. You may as well not bother to make a report at all. Instead, a note such as "Place 30 cy conc. footing along north wall between col. lines H & I, and 22 cy conc. foundation wall along north wall between col. lines F & H," will more positively identify the actual work performed.

   Treatment of your remarks in this manner is probably the most important consideration of all in the correct preparation of the Daily Field Report.
8. Identify all sources of information positively. Any information other than your own observations must have its source(s) identified. Refer to companies and *name names.*
9. Include a Field Report for *every* workday. If, for any reason, no work is performed on a specific day, fill out the day's report, indicating "no work." Be sure to include a statement of the reason for the condition, such as "still waiting for Walkway Change Order (033) approval before any work can proceed," or "HVAC Subcontractor no-show."
10. Include a report for every non-workday. Provide a report to account for each weekend and holiday. If no work was done, days (such as Saturday and Sunday) may be combined in a single report.

### 5.9.6 Report Information Guidelines

1. *Title Box.* Indicate project name, Company job number, project location, and the name of the individual completing the report.
2. *Date / Page.* There will always be a minimum of two pages available for a day's report. The first sheet is the identification and administrative portion, with at least one Field Work Report Form included to describe the physical work. As many Field Work Report sheets are to be used as necessary for a comprehensive report. Never use the back of any report form.
3. *Weather*
   - *a.* Include a short remark on the typical condition of the day (cloudy, rain, heavy snow, clear, etc.).
   - *b.* Record the temperature in degrees F at the beginning, middle, and end of the day.
4. *Staff (S / V).* List all Company staff on the site that day. Begin with all permanent staff, indicating "S" (site) after their names. Follow with Company personnel visiting the site that day, noting "V" after their names.
5. *Equipment (C / S)*
   - *a.* List all equipment on the jobsite that day. Indicate:
     - (1) Quantity and type
     - (2) Whether it is a Company piece (C) or the responsibility of a particular Subcontractor (S); name the Subcontractor.
     - (3) Include every piece every day.
   - *b.* Describe the work performed with a brief note that will be presumably elaborated upon in the Field Work Report portion of the report. Include the estimated number of hours worked, or check whether the equipment was idle that day.
   - *c.* If it arrived or departed that day, indicate the time in the area provided. Otherwise leave the lines blank.
6. *Visitors / Conversations / Meetings / Photos*
   Use the jobsite Telephone Log of Section 5.8 as a supplement.
   - *a.* List every visitor to the jobsite that day, and include a brief remark as to the reason for their visit.
   - *b.* Note every relevant conversation with anyone at the site or on the phone.
   - *c.* If the comments, conclusions, direction, etc., are short enough to be included completely here, do so. If not, identify the fact and refer to the appropriate detailed record (job meeting minutes, Company record of Phone Conversation/Meeting Form, a confirming memo or fax, etc.).
   - *d.* Use this area to summarize new or continuing problems, or to indicate any miscellaneous remarks that are relevant to the work.
   - *e.* Indicate if regular or special photos have been taken and the corresponding reason.
7. *Required Materials / Information*
   - *a.* Summarize your requests both to Subcontractors and Suppliers and to Company personnel for any information or materials, and the date promised. Again, name names.
   - *b.* This is *not* to be considered your complete documentation effort. It is only a summary note, and should only be there to catalog your more elaborate noti-

fications, memos, faxes, etc., that have been properly supported, documented, and distributed in strict accordance with the requirements of those respective portions of the Manual.

8. *Field Work Report* (*Page 2+*)
   *a.* Record all on-site activity in accordance with the recommendations of Section 5.9.5, *Report Preparation Guidelines.*
   *b.* Start each entry with a Subcontractor's company name or the Company name, underlined.
   *c.* Indicate the number of each classification of worker, separating foremen. Indicate the correct EEO code for each individual.
   *d.* Include complete and comprehensive descriptions of specific work performed for each company and for each group of workers. Include enough information so that anyone who has not been on the site that day can understand and identify precisely what went on.
   *e.* On the right side of the form indicate if the work described is the subject of a change order, proposed change, possible claim, done under protest, etc. Include the Company file number and Owner change order number whenever there is one available. If there is no Company number available, contact the Project Engineer and secure a new number assignment immediately.
9. *Distribution.* Attach any Subcontractor or Owner reports if they are available, and either forward to the office with your payroll records or mail in daily, as determined by the Project Manager.

Follow the completed example of the Sample Daily Field Report Form in Section 5.9.7 as further guidance in the preparation of these critical project documents. A blank form is given in Section 5.9.8.

### 5.9.7 Sample Daily Field Report Form—Completed Example *(page 5.36)*

### 5.9.8 Sample Daily Field Report Form—Blank Form *(page 5.38)*

## 5.10 Equipment Use Release Form

### 5.10.1 General

The Equipment Use Release Form is to be used in all cases where any Subcontractor is allowed to use any Company equipment or facility. Its purposes are to:

1. Describe the equipment authorized to be used
2. Confirm the specific use allowed, including scope, time frame(s), and location(s)
3. Confirm the acknowledgment or responsibility for all costs, effects, damages, etc., resulting from the use of the equipment by non-Company personnel

## 5.9.7
## Sample Daily Field Report
## Page 1 of 2
## (Completed Example)

Project: FIREHOUSE ADDITION No: 9424 Date: MAY 15, 1994 Page ___ of 2

Location: NEW CITY, CT Weather: CLEAR / BREEZY

Superintendent: D. JEFFKO Temp: 8AM 71° 1PM 79° 4PM 77°

STAFF

| Name | Classification | Name | Classification |
|---|---|---|---|
| D. JEFFKO | SUPER | | |
| M. LEONARDO | P.E. | | |
| | | | |
| | | | |
| | | | |

EQUIPMENT

| Quant | Co/Sub | Type/Size | Work/Idle | Work Performed | Arrive | Depart |
|---|---|---|---|---|---|---|
| 1 | S | CJ7 DOZER | / ✓ | | | |
| 1 | S | 920 LOADER | / ✓ | | | |
| 1 | S | 770 HIGH-LIFT | ✓ / | EXTER. BRICK VENEER | | |
| | | | / | | | |
| | | | / | | | |
| | | | / | | | |
| | | | / | | | |
| | | | / | | | |
| | | | / | | | |
| | | | / | | | |

VISITORS / CONVERSATIONS / MEETINGS / SAFETY REVIEWS

D.J. 9:20 AM TELECON W/RAY SMITH (ARCHITECT); RAY EXPECTS TO HAVE A-LINE CLARIFICATION (ANSWER TO RFI # 12) BY 5-16-94

J. CARRIGAN (OWNER REP) CONDUCTED GENERAL SITE REVIEW BETWEEN 11:10 AM & 12:00 NOON

R. KROY (KROY ROOFING) STOPPED BY AT 2:40 PM TO CONFIRM THAT SITE IS READY FOR KROY'S MAT. DELIVERIES ON 5-18-94

REQUIRED MATERIAL / INFORMATION

| Item | Requested From | Company | Promised By |
|---|---|---|---|
| RFI #12 ANSWER | R. SMITH | A&S ARCHITECTS | 5-16-94 |
| TEMP. PROTECT. MAT'L FOR H.M. | P. DAVIES | D&V MASONRY | 5-16-94 |
| | | | |
| | | | |
| | | | |
| | | | |

Ray Day
Project Manager (Signature)

David Jeffko
Superintendent (Signature)

## 5.9.7
## Sample Daily Field Report
## Page 2 of 2
## (Completed Example)

Page 2 of 2

FIELD WORK REPORT

Project: FIREHOUSE ADDITION # 9424 Date: MAY 15, 1994

| # Workers | Classification | EEO Code | Complete Description and Location of Work | CO# |
|---|---|---|---|---|
| | | | STATESIDE FRAMING | |
| 3 | CARP. | WM | CONTINUE FRAMING EXTER 16 GA. STL. STUD & | |
| 1 | CARP. FOREMAN | BM | GYP SHEATHING @ SOUTH ELEVATION | |
| | | | | |
| | | | HEAT & AC, INC. | |
| 1 | FOREMAN | WM | · COMPLETE CHANGES TO F-C UNITS | #1 |
| 1 | PLUMBER | WM | · CONTINUE DUCT BRANCHES IN ROOMS | |
| 1 | " | WF | 219, 220, 224, & 227 | |
| | | | | |
| | | | SHORE ELECTRIC | |
| 1 | FOREMAN | HM | · COMPLETE BRACH WIRING @ 2ND FLOOR | |
| 1 | ELECTRICIAN | BM | · CONTINUE LIGHT FIXTURE INSTALLATION @ 1ST FLOOR | |
| 1 | " | WM | ROOMS 112, 112A, & 113 | |
| | | | | |
| | | | D & W MASONRY | |
| 1 | FOREMAN | WM | · CONTINUE BRICK VENEER AT NORTH & EAST | |
| 2 | MASONS | BM | EXTERIOR ELEVATION | |
| 2 | MASONS | WM | · COMPLETE 2ND FLOOR (FINAL) ELEV. SHAFT | |
| 2 | LABORERS | WM | | |
| 17 | TOTAL FORCE | | | |

# 5.9.8
# Sample Daily Field Report
# Page 1 of 2

Project:________________________ No:__________ Date:____________________ Page___ of___
Location:________________________________ Weather:________________________________
Superintendent:____________________________ Temp: 8AM ______ 1PM ______ 4PM ______

STAFF

| Name | Classification | Name | Classification |
|---|---|---|---|
| | | | |
| | | | |
| | | | |
| | | | |
| | | | |

EQUIPMENT

| Quant | Co/Sub | Type/Size | Work/Idle | Work Performed | Arrive | Depart |
|---|---|---|---|---|---|---|
| | | | / | | | |
| | | | / | | | |
| | | | / | | | |
| | | | / | | | |
| | | | / | | | |
| | | | / | | | |
| | | | / | | | |
| | | | / | | | |
| | | | / | | | |
| | | | / | | | |
| | | | / | | | |

VISITORS / CONVERSATIONS / MEETINGS / SAFETY REVIEWS

REQUIRED MATERIAL / INFORMATION

| Item | Requested From | Company | Promised By |
|---|---|---|---|
| | | | |
| | | | |
| | | | |
| | | | |
| | | | |
| | | | |
| | | | |

Project Manager (Signature ) ____________________ Superintendent (Signature) ____________________

## 5.9.8
## Sample Daily Field Report
## Page 2 of 2

Page _____ of _______

FIELD WORK REPORT

Project:_________________________________ #_______________ Date:__________________________

| # Workers | Classification | EEO Code | Complete Description and Location of Work | CO# |
|---|---|---|---|---|
| | | | | |

4. Secure the legal release of Company liability, and the user's agreement to hold the Company harmless for all loss and damage, including property damage, accident, dismemberment, death, or any result whatsoever

*Note:* Check with your own attorney to confirm the specific language necessary and/or allowed in your area. The form provided in Section 5.10.3 is an example only, and is not to be considered as legal advice or the specific legal document to be used.

### 5.10.2 Use and Procedure

1. The form's most common application is in the case of scaffolding provided by the Company for use by several trades. It is, however, to be used in any case in which any non-Company person or company is allowed to use any Company equipment or facility.
2. The form must be filled out completely, and in duplicate. Each copy must be executed by a person authorized by their company to do so.
3. Return one executed copy to the home office to be filed in the respective Sub-Vendor's bid package file; retain the other at the jobsite.
4. Have a new form prepared and executed for any use and/or time frame not specifically described on any previous form.

### 5.10.3 Sample Equipment Use Release Form

## 5.11 Preconstruction Survey

### 5.11.1 General

The preconstruction survey is the documentation effort that establishes to the best degree possible the precise existing conditions prior the the start of *any* work at the site. It should begin prior to actual mobilization, and continue as appropriate throughout the mobilization period. When the initial preconstruction survey effort is completed, it may become necessary at various points of construction (particularly in renovation projects) to supplement the survey with new information for areas that are becoming available.

The survey effort will include:

1. Preconstruction photographs
2. Preconstruction video
3. General verification of existing site information, both for properties directly adjacent to the site, and for the site itself within the contract limit lines.

These processes are closely related to and should be coordinated with Section 5.16, *Construction Photographs,* and Section 5.12, *Field Engineering, Layout, and Survey Control.*

## 5.10.3
## Sample Equipment Use Release Form

Date: ____________________

Project: ______________________________ #__________
Location: ______________________________

Company: ____________________
____________________
____________________

Authorized Representative: ____________________

Equipment to be used by personnel of the company indicated above
(Completely describe all equipment, including specific locations at all areas of the jobsite and summary descriptions of purposes):

______________________________
______________________________
______________________________
______________________________
______________________________
______________________________
______________________________

Date(s) and time(s) to be used (Include every possible instance):

______________________________
______________________________
______________________________
______________________________
______________________________
______________________________
______________________________

As an authorized representative of ____________________ (Company indicated above), we agree to use only the equipment specifically described above in the locations indicated, on the dates listed, and within the hours indicated.

We agree to strictly abide by all ____________________ Company, project, safety, government, rules and regulations, and to comply with all requirements of OSHA and all other entities having jurisdiction over the work.

We agree to be completely responsible for any and all direct, indirect, and consequential damages caused by any act, omission or negligence on the part of ____________________ (Company indicated above), and to hold ____________________ Company, including its officers, agents, servants, and employees harmless from any damages of any nature or kind whatsoever.

AGREED TO BY:

Company: ____________________
Authorized Individual (Signature): ____________________
Date: ____________________

Executed in the Presence of (Signature): ____________________

## 5.11.2 Preconstruction Photographs

Ideally the preconstruction photo set should be taken even before the field office has been mobilized. Every area of the site, every adjacent property, and all surrounding private and public properties must be photographed extensively in every detail. Procedures that will efficiently accomplish this include the following:

1. *Use a 35-mm camera of good quality.*
   - This is necessary in order to preserve the ability to select specific photos later, and to maintain acceptable quality in enlargements.
   - Autofocus is available on many cameras with little or no additional cost. It is not required, but highly desirable.
   - Many cameras have optional camera backs available that will stamp the specific photo optically with the date. These are reasonably priced and should be used if possible.
   - Autoflash systems are available for little or no additional cost on some cameras, or at least at little additional cost compared to the price of a conventional flash. Although not required, it is a great feature, which saves time while guaranteeing correct exposure.
2. *Organize your approach.* The photo sets will be divided between:
   - The site itself (within the contract limit lines)
   - *All* properties immediately adjacent to the site
   - Approach routes to the site
3. *Keep each area as a distinct set.*
   - Start in a logical place and proceed methodically through the entire area. Do the entire site first, then photograph every property adjacent to the site. Pay particular attention to areas along the property lines, and to all physical construction (buildings, parking areas, fences, etc.).
   - Make particular notes of all existing damage—cracks, settlement, damaged surface finishes, etc.
   - Refer to the site utilization program of Section 5.4 to determine the general routes of all construction and worker vehicle traffic into and out of the site. Include photo sets of the approaches of these routes into the site.
4. *Secure permission for complete photo sets.* Whenever there is any existing building or structure that may be affected by any construction operation such as blasting, pile driving, or dewatering, contact the owner of such property in an attempt to secure permission for a detailed preconstruction photo survey of the complete property. If such permission is obtained, photosurvey the entire premises. If permission is not obtained, photograph as many views and features as are possible from within project property lines. In either event, pay particular attention to all existing apparent damage, such as foundation cracks, badly maintained landscaping and grounds, and settlement of older structures.
5. *When in doubt, shoot first and ask questions later.* Film is the least expensive part of project documentation that will have the most dramatic effect on settling disputes and saving big dollars. *Never* skimp on film, even with the noble intention of saving a few dollars in the process. Keep your priority on ensuring the most complete, thorough, and comprehensive photo survey possible.
6. *Identify and date the survey.* Note the survey on appropriate Daily Field Reports and include a copy of the reports in the survey record file. Include any narrative or other appropriate description to make the file complete.

7. *Have all photographs developed immediately.* Do not leave undeveloped film as the photo record. Not only is it cumbersome to deal with and difficult to file properly, but you will not be able to research the file. There is also no guarantee that the photos all came out (or that they are actually not of some stranger's vacation…).
8. *File correctly.* Forward all developed photos with the field reports and other relevant documentation to be filed permanently at the home office.

### 5.11.3 Preconstruction Video

It is becoming increasingly more desirable to supplement the preconstruction photo survey with a preconstruction video. Even if the Company does not own its own video camera, they are so readily available from individuals or by rental that there is really no excuse not to use one.

The preconstruction video is *not* a substitute for the preconstruction photos. It is a supplement. Difficulties with the video lie principally in the limited ability to select, reproduce, and enlarge specific photographs for review and/or demonstration.

The major advantage of the video is that the record picks up so much more on the tape than the photographer is actually observing at the time. It ties the survey together to clearly demonstrate relationships of each area.

Follow the guidelines of the preconstruction photographs of Section 5.11.2 to organize your approach and take the video. Additional recommendations to help an acceptable production of the video include:

1. Walk slowly and hold the camera *steady*.
2. Proceed along a planned route. Pan the camera very slowly.
3. Narrate.
   - Start the tape with the complete project identification, date, photographer's name, and any other relevant information.
   - Describe each view. Try to do so in a manner that would allow a person unfamiliar with the site to locate the area, identify the point of view, and understand what he or she is observing.
4. Return the completed video to the home office to supplement the regular preconstruction photograph record file.

### 5.11.4 General Verification of Existing Site Information

The final component of the preconstruction survey is a general review of the various conditions of the property as compared to those actually represented in the contract documents and/or observable during a prebid site review. This review is not complicated, but can disclose a surprising amount of issues and questions before the first cubic yard of material is removed. It will protect you, the Company, and even Subcontractors from exposures ranging from the repair of simple damages that existed prior to construction start to problems as significant as removing hundreds or even thousands of cubic yards of earth off the site.

The considerations are again divided between properties adjacent to the construction area and the construction site itself within the contract limit lines.

### 5.11.5 Adjacent Properties

The characteristics of the properties immediately adjacent to the construction site that are or have the potential to become significant may not be easily apparent. For many of them to become visible, it may boil down to being in the right place at the right time—to be in a specific area just as the problem is occurring.

The kinds of problems that will fall in this category may have been present during the bidding process, but because of their nature have been effectively hidden during the prebid site investigation (and even from the design process itself). The Owner may have been aware of the condition, but not aware of any possible effect on construction, or there may be other reasons why their inclusion in the bid documents may have been considered unnecessary or inadvisable at the time. On the other hand, even the Owner and the design professionals may be completely justified in not being aware of the adjacent properties' characteristics due to their cryptic nature.

Examples of the kinds of conditions that have the potential of impacting construction operations seriously include:

1. A seasonal watercourse that drains several acres of property directly into your footing excavation
2. Heavy traffic that restricts perimeter mobility during key hours of the day
3. Construction on an adjacent site that no one was properly aware of, which had begun before this one, presenting unanticipated problems for your own operation (shoring, dewatering, access, staging, etc.)
4. Undisclosed unusual ordinances or other regulations limiting noise during regular working hours

It can take a great deal of imagination to actually identify these kinds of conditions early on. They are the most difficult of all conditions to spot before they become painfully apparent. The purpose of their identification here, then, is first to make the Superintendent aware of the survey category. From this awareness the Superintendent must make an effort to:

1. Do his or her best to actually identify these conditions early as best possible
2. Be constantly aware of the potential for these kinds of conditions; and to be quick to identify them for what they are if any should appear at any point during construction

Whenever such an actual or possible condition is observed, the Project Manager must immediately be notified. Fast action will become critical to:

1. Confirm the actual circumstance
2. Prepare and deliver all appropriate notifications to the Owner, design professionals, and all potentially affected Subcontractors
3. Develop at least one (or preferably optional) corrective action plan(s) to be presented to the construction team for consideration as to the most desirable or sensible way to proceed, considering all general cost and schedule implications

4. When the best plan is decided upon, develop a complete analysis of all costs and schedule impacts to be used for ultimate presentation to the Owner in order to secure appropriate related contract modifications of cost and time

Consider the Pass-Through Clause of Section 3.3.9. Know in advance whether it is the Company's immediate responsibility to deal with the specific problem (heavy traffic, for example) or whether the problem really belongs to a particular Subcontractor (water draining into an excavation).

If it is determined to be the responsibility of a Subcontractor, advise him or her of the condition for their own determination of how they think they should proceed. Always be aware of the Company's contractual responsibilities, both to the Owner and to respective Subcontractors, and have a clear understanding of the relationships before proceeding.

If it is determined to be a Company issue, consider the Sample Letter to Owner Regarding Unanticipated Effects of Adjacent Properties of Section 5.11.6.

### 5.11.6 Sample Letter to Owner Regarding Unanticipated Effects of Adjacent Properties *(page 5.46)*

The Sample Letter to Owner Regarding Unanticipated Effects of Adjacent Properties that follows is an example of the kind of notification to be sent to the Owner upon the discovery of some condition on an adjacent property that:

1. Did not exist or was not apparent at the time of bid
2. Affects construction directly in an adverse way by delaying the project and/or increasing cost

It is intended, first, to guarantee compliance with any contractual notification requirement by confirming such notice immediately upon the discovery of even a potential condition and, second, to push the resolution procedure into motion.

Before using the letter or anything similar, consider the Pass-Through Clause of Section 3.3.9 to make sure that dealing with the condition is really not a Subcontractor's obligation (instead of the Company's).

### 5.11.7 Verification of Grades, Elevations, and Contours

It is ideal to begin this component of the preconstruction survey prior to mobilization, but as a practical matter, the procedure will continue into the actual mobilization period and even into the actual site start-up period. In any event, it must be conducted prior to the start of any work in the affected areas.

Before any existing site conditions are disturbed, the actual state of affairs must be verified physically and compared to the conditions represented in the contract documents. Two major categories of site information to be verified include:

1. Grades, elevations, and contours (discussed here)
2. Existing site constructions (discussed in Section 5.11.9)

## 5.11.6
## Sample Letter to the Owner
## Regarding Unanticipated Effects of Adjacent Properties

**Letterhead**

(Date)

Confirmation of Fax
Fax#:____________________

To : (Owner)

RE: (Project Description)
(Company Project #)

SUBJ: Unanticipated Effects of Adjacent Property:
Watercourse at North Property Line

Mr. (Ms.)

On (date); one day after heavy rain, it was observed that the several acres directly north of the project property line (contract limit line) drain into a natural swale. This swale creates a watercourse that directs the flow of the surface drainage of that property directly into the footing excavation along the north property line.

We have begun dewatering, but significant water continues to drain into the excavation even two days after the rain. Our growing concern is that we are moving into April; characteristically the rainy season. We therefore now unfortunately anticipate that this will become a chronic and growing problem.

Because the project was bid in August of last year, the natural watercourse could not be apparant to anyone during any pre-bid site investigation, and no such water problem is indicated anywhere in the contract documents. We are accordingly advising you that the project is being impacted in a manner that will affect both cost and time.

As time is of the essence, we request that you immediately have your design professionals review the condition, and advise us on the specific way in which you wish us to proceed. When that information is received, we will submit the complete proposal for the work and its effects.

Very truly yours,

COMPANY

Project Manager

cc: Jobsite
Architect
Engineer
File: Site, Change File ( ), CF

**Grades, elevations, and contours.** The contours of the new site design are normally prepared by either a Registered Land Surveyor or a Professional Engineer. At times, existing data are verified to some extent before the new design proceeds, but at times they are not. In these instances, grades and contours may, for example, have been lifted from existing land surveys or prior As-Built Drawings.

During the bid preparation, the contractor has no choice but to rely on the accuracy of this design information. Accordingly it affects the estimates for nearly everything on the site, including, for example:

- Earthwork cuts and fills
- Demolitions and alterations
- Trench excavation lengths, widths, and depths
- Sequence of site activities
- Planned locations of material stockpiles, staging areas, and field offices
- Provisions for temporary power, phone, and water
- Surface water control
- Time and extent of road excavations and tie-ins
- Pavement cutting and patching
- Excavation shoring
- Treatment of mass and trench rock

#### Action

1. Before any work is done, the photographs and possible video will have been made in accordance with the recommendations of Sections 5.11.2 and 5.11.3.
2. Spot-check existing grades at several easily identifiable locations. If these checks prove to be accurate, note them on the As-Built Drawings, and continue to the point where you are reasonably confident in the remaining information. If no significant discrepancies or errors are discovered at this step, the work should be allowed to proceed.
3. If *any* discrepancies are discovered, this should be your signal to step up the level of detail of your verification efforts. Identify a key area of the site, for example, and conduct a detailed check of the contour information.
4. If this detailed check confirms multiple or significant discrepancies, notify the Project Manager immediately. Reconsider the validity of your own information, and determine if those checks themselves should be confirmed before letting the world know of the problem. Time is truly of the essence.
5. If the potential extent of the contour discrepancy is large, consider immediately securing the services of a Registered Land Surveyor or Professional Engineer to substantiate and detail the precise actual condition. This decision must be made with the Project Manager.
6. Whether or not a Land Surveyor or a Professional Engineer becomes involved at this point, the Owner and the design professionals must be notified at the earliest moment when the discrepancies are positively confirmed. Consider using the *Sample Letter to Owner Regarding Discrepancies in Existing Grades and Elevations* of Section 5.11.8 as an example of the kind of required notification.

7. Depending upon the extent of the discrepancy and the potential value of corrective work and/or other project effects, stopping work in the affected area might be considered. This, however, is a very serious action, and should only be considered if for some reason preservation of the area is necessary for prompt and/or equitable resolution of the problem. *No* stop-work action may be taken without confirmation by Company senior management as to the appropriate procedure.
8. Rather than the Company assuming such profound responsibility for decision to stop, proceed, or take any other action, it will likely be more advisable, as part of the Owner notification, to *request specific direction* as to whether to stop or to proceed. Request such information by a specific date, after which other action may be necessary to mitigate the complete effects on the project.

### 5.11.8 Sample Letter to Owner Regarding Discrepancies in Existing Grades and Elevations *(page 5.49)*

The Sample Letter to Owner Regarding Discrepancies in Existing Grades and Elevations that follows is an example of the notification discussed in step 6 of the previous section and should be sent by the Project Manager. Specifically, it notifies the Owner that:

1. It has been confirmed that discrepancies exist in a specific area of the site.
2. Research is continuing, and engineering data will be provided.
3. Complete total costs, effects on activity sequences and durations, and overall job impact will be calculated. When all this information is known, it will be packaged and submitted for payment.
4. It *requests direction* from the Owner to either stop or continue work. The direction is requested by a specific date.
5. Rights are reserved to claim additional costs for effects that are unforeseen at this time.

*Strongly* consider the inclusion of the last sentence of the letter at this point. Be aware that this letter will be among the very first of such communications to the Owner, and as such will play a major role in setting the tone for the project. It may instead be more advisable to leave this last phrase off and save it for the second letter if the Owner does not give you an appropriate response by the required date. Let someone else be the first to shoot him- or herself in the foot.

### 5.11.9 Verifications of Existing Site Constructions

The considerations for the verification of existing site constructions are identical to those for grades, elevations, and contours of Section 5.11.7. For the same reasons as listed before, engineering data relative to the actual configuration of existing site constructions (as opposed to those configurations included in the contract documents) may not have been properly verified before the new designs were prepared. Problems could, for example, exist relating to situations such as the following:

## 5.11.8
## Sample Letter to the Owner
## Regarding Discrepancies in Existing Grades and Elevations

**Letterhead**

(Date)

To: (Owner)

RE: (Project)
(Company Project #)

SUBJ: Discrepancies in Existing Grades and Elevations

Mr. (Ms) :

It has been discovered that the existing site grades and elevations differ materially from those shown on Drawing L-1. Per my conversation with your (Name) on (Date), we have retained a registered land surveyor to prepare a complete analysis. We expect the information to be complete by (Date) and will immediately forward it to you.

Upon completion of the survey, an analysis of the changed work, including any effects on activity durations and sequences will be completed, and you will be advised of any changes in cost and/or construction program.

Your direction is therefore requested to either stop work, or to proceed with any related additional work. Your response is required by (Date) in order to minimize interferences caused by these conditions.

The resolution of this problem will profoundly affect our ability to properly proceed with the work. The utmost urgency is therefore stressed. We accordingly must reserve all rights to claim all additional costs that are unforeseen at this time.

Very truly yours,

COMPANY

Project Manager

cc: Architect
Jobsite
File: Site, Change File ( ), CF

- Manhole locations and storm and sanitary line invert elevations originally provided by the City Engineer's office or other authority may be based upon aging as-built information that is not accurate.
- Telephone and power line locations were verified with the respective utility companies; or maybe they haven't been.
- Transformer pad locations and configurations have or have not been coordinated with the utility company.

#### Action

1. Again, photographs and a possible video should have already been taken per Sections 5.11.2 and 5.11.3.
2. Check the locations of manholes, markers, or any available indicator to help verify the locations of existing telephone, water, sewer, and gas lines and fuel tanks.
3. Open a few manholes. Spot-check actual invert elevations and compare them with those shown on the plans.
4. Check the locations of telephone poles, street signs, pole guys, and any other constructions. Confirm that they do not interfere with new roads, walks, pavement, excavations, or other site improvements.
5. Check the actual horizontal distances among telephone poles, light poles, manholes, drainage structures, and so on. Compare them to those indicated on the plans.
6. Significant and/or numerous discrepancies may be a strong indicator of the overall lack of quality of the plans themselves, and may thereby strongly suggest additional detailed investigations. In this instance, verify your own information and notify the Project Manager immediately.
7. As with the condition for discrepancies in site contours and elevations, the Owner must be notified immediately upon your confirmation of the accuracy of your own information.

It is important to be aware of the potential effects on the project, which may be limited and immediately apparent:

> **Example:** CB #2 needs to be raised 1 ft to make the drainage work…,

Or the effects may in the final analysis turn out to be profound:

> **Example:** When CB #2 is raised, CB #1 no longer works; when that's raised, the parking lot must be picked up, affecting adjacent grades, that may affect permit considerations…

The potential for these kinds of extended effects may or may not even be apparent at the early stages of discovery and review.

It is therefore crucial that all actions, notifications, and considerations be taken with utmost care to avoid assuming unnecessarily the responsibility for the ripple effects of subsequent necessary design changes. It will *only* be the Company's job to notify the Owner of the problem, leaving complete determination of the solution with the Owner (and with the Owner's engineer's professional liability insurance).

Consider using the *Sample Letters to Owner Regarding Changed Site Conditions* of Sections 5.11.10 and 5.11.11 as examples of such a notification:

- Letter #1 can be used in the instance where a changed condition is very easily identifiable as one that can be corrected without significant effect on surrounding or contiguous construction.
- Letter #2 can be used in the instance where a changed condition has the possibility for significant effect on surrounding or contiguous construction.

### 5.11.10 Sample Letter #1 to Owner Regarding Changed Site Conditions—Simple Condition *(page 5.52)*

The Sample Letter #1 to Owner Regarding Changed Site Conditions that follows is an example of the first of such notifications discussed in step 7 of Section 5.11.9 and should be sent by the Project Manager. Its use should be considered in those cases where it is easily apparent that correction of the respective situation has little to no effect on surrounding construction. (If this is not the case, consider Letter #2.) Specifically, it notifies the Owner that:

1. It has been confirmed that discrepancies exist between the actual conditions of certain items and those conditions indicated in the contract.
2. All substantiating calculations are complete, and the costs are included in the attached change order proposal.
3. Owner approval is required before the changed work can proceed.
4. No responsibility is assumed for the new design; approval is therefore construed to confirm that the project design engineer has incorporated the new design into the project and as such is covered by his or her own design liability.
5. Rights are reserved to claim costs and damages that are unforeseen at this time.

As with the other letters of this section, consider this last phrase carefully. Be aware that this letter will be among the very first communications between the Company and the Owner, and as such will play a major role in setting the tone for the entire project. It may instead be more advisable to leave the last phrase off, saving it for the next letter if the Owner's actions become inappropriate.

### 5.11.11 Sample Letter #2 to Owner Regarding Changed Site Conditions—Complex Condition *(page 5.53)*

The Sample Letter #2 to Owner Regarding Changed Site Conditions that follows is an example of the second of such notifications discussed in step 7 of Section 5.11.9 and should be sent by the Project Manager. Its use should be considered in those cases where it is a possibility that correction of the respective immediate situation definitely has or may have *any degree* of effect on surrounding construction. (If this is not the case, consider Letter #1.) Specifically, it notifies the Owner that:

1. It has been confirmed that discrepancies exist between the actual conditions of certain items and those conditions indicated in the contract.
2. The conditions affect surrounding work, and will likely cause some degree of redesign in contiguous construction.

## 5.11.10
## Sample Letter #1 to the Owner
## Regarding Changed Site Conditions: Simple Condition

| Letterhead |
|---|

(Date)

To: (Owner)

RE: (Project)
(Company Project #)

SUBJ: Changed Site Condition due to
Discrepancies in Existing Elevations

Mr. (Ms) :

Drawing SU-1 indicates the pipe invert at Existing MH #2 to be at 35.5'. The actual invert elevation is 25.5'. The additional 10' in depth of the invert places the new pipe into rock.

The corresponding increase in the contract price to cover the cost for the additional work is ($ ), computed in accordance with the procedures for Earth and Rock Excavation General Conditions Article 4. The details of this calculation are attached.

Is seems that simply lowering the pipe to meet the bottom of MH #2 will not affect surrounding construction. We cannot, however, assume design responsibility for the design change. We therefor offer this solution as suggestion only, to be considered by your engineer; the final design must be provided by your design professionals. Any change between the final design and those conditions represented here will cause the cost proposal to change accordingly.

Your confirmation of the design and approval of the change is required by (Date) in order to allow the work to proceed without additional interruption. All rights are reserved to claim all additional costs that are unforeseen at this time.

Very truly yours,

COMPANY

Project Manager

cc: Architect
Jobsite
File: Site, Change File ( ), CF

## 5.11.11
## Sample Letter #2 to the Owner
## Regarding Changed Site Conditions: Complex Condition

**Letterhead**

(Date)

To: (Owner)

RE: (Project)
(Company Project #)

SUBJ: Changed Site Condition due to
Discrepancies in Existing Elevations

Mr. (Ms) :

Drawing SU-1 indicates the pipe invert at Existing MH #2 to be at 25.5'. The actual invert elevation is 35.5'. The discrepancy may be the result of an unfortunate typographical or transpositional error.

Simply raising the invert is not possible, as it will cause all surrounding drainage based upon the incorrect invert to fail. The drainage throughout the entire area must therefore be redesigned, hopefully in such a way that will minimize resulting effects on the current design for the new parking area.

This is a very serious problem that has stopped all work throughout the entire area of the site as of (Date). Utmost urgency is stressed to resolve the redesign and to approve the subsequent change order as soon as possible in order to mitigate the potentially extreme consequences to the project. Time is truly of the essence.

We are available to meet at your convenience to assist in the expedient resolution of this problem in any way that you may require. When the redesign is finalized, you will be advised of the anticipated effects on the project's cost and time.

Very truly yours,

COMPANY

Project Manager

cc: Architect
Jobsite
File: Site, Change File ( ), CF

3. Work in the affected areas cannot proceed until the completed redesign is confirmed and the associated change order is prepared, reviewed, and approved.
4. The project is dramatically impacted. Because the new design does not yet exist, and the change order has not been finalized, there is no way to anticipate the net effect on the project. Accordingly, when all these resolutions have been finalized, the complete effect on the project will be determined, along with the resulting effects on cost and time.
5. Rights are reserved to claim costs and damages that are unforeseen at this time.

As with the other letters of this section, consider this last phrase carefully. Be aware that this letter will be among the very first communications between the Company and the Owner, and as such will play a major role in setting the tone for the entire project. It may instead be more advisable to leave the last phrase off, saving it for the next letter if the Owner's actions become inappropriate.

## 5.12 Field Engineering, Layout, and Survey Control

### 5.12.1 Responsibility, Organization, and Description of Work

The Superintendent must first be aware of the direct responsibility to lay out the work physically. This begins with determining the responsibility to identify and to physically establish the starting baselines and benchmark, and follows through the work of each respective bid package.

The categories of field engineering are therefore divided between the physical layout and the coordination of:

1. The overall building and site
2. The work directly performed by the Company
3. The work performed by Subcontractors or Trade Contractors

As an industry, there is a pronounced trend among General Contractors in the movement of their company structures toward larger percentages of subcontracted work, until the project is effectively 100 percent subcontracted. This form of management has distinct advantages in cost performance and management efficiency. There are, however, obvious and subtle approaches that must be taken in many areas—field engineering and coordination of work among them—if these advantages are to be realized, and not simply an inordinate amount of risk assumed on the part of the Company.

There are clear, distinct responsibilities for field engineering, layout, and coordination of the various parts of the work that are very well represented in the respective general contract and subcontract agreements. The successful Superintendent will not, however, assume that every party to the contract:

- Is aware of its responsibilities
- Intends to follow through on them
- Is capable and qualified to complete them in every respect

- Is competent in its efforts to coordinate with the layout work of other trades

The Superintendent must:

- Know in every case what the specific responsibilities of each party to the contract are
- Be able to immediately recognize the presence or lack of intention, competence, and qualifications
- Constantly devote an *appropriate* amount of attention to policing the effort and *verifying the information*
- Take immediate, appropriate action when small or serious deficiencies are observed

### 5.12.2 Baselines and Benchmark

It is most common for the Owner to be responsible for providing a minimum of two baselines and a benchmark from which the Contractors will lay out the remainder of the work.

Know what the language of the general contract specifically requires. It is not usually enough for the Owner's representative to point out the locations of the baselines and benchmark on the plans and leave you with the responsibility to establish them in the field. It is more likely to be the Owner's responsibility to *physically establish* these fundamental construction starting points.

This idea is not so much a cost issue as a major liability one. Because the baselines and benchmark are the starting data against which the entire project will be built, any error at all in this initial setup can cause problems with almost every conceivable part of the construction. While it is true that the difficulty in establishing these points is minimal, and in most cases the probability of error is for practical purposes normally very low, the potential size of any problem if there is an error is just too great.

If, as is customary, it is clearly the Owner's responsibility, but the Owner for any reason should refuse, immediately consult with the Project Manager to confirm contractual responsibility and force appropriate compliance. This is not a negotiable point with the Owner.

### 5.12.3 Site and Building Layout and Procedure

Although the Site and Concrete Subcontractors are technically responsible (as are the other Subcontractors and trade Contractors) to lay out their own work (refer to the Pass-Through Clause of Section 3.3.9), the effects on the project resulting from even a small error in layout in these areas can be too catastrophic to leave this function completely up to someone else. These, then, are the two bid packages in which exceptions to the Pass-Through Clause must be made to a significant degree.

In the past it has been most common for the General Contractor to lay out the work with its own forces. Too many of these Contractors, however, have compiled their stories of foundations, parking areas, and other structures being in the wrong location. Be sure that *you are* the Superintendent who through consistent, proper attention doesn't have one of these stories of your own.

The procedure, then, to provide the building and site layout is as follows:

1. Ensure that the Owner has physically established a minimum of two baselines and one benchmark at the site.
2. Arrange to have a Professional Engineer or a Registered Land Surveyor perform all point establishments as described in the remainder of this section.
   *a.* These must be certified professionals with appropriate professional liability insurance.
   *b.* There are no exceptions. General site and building layouts must not be performed by anyone without appropriate certification and liability insurance.
   *c.* Prior to beginning such a layout, have the survey company and/or individuals submit copies of their professional liability insurance policies to the Project Manager for confirmation.
3. Arrange a coordination meeting with the survey crew, the Concrete Subcontractor, the Site Subcontractor, and any other Subcontractor or trade Contractor immediately affected by the location of survey information (stakes, points, etc.).
   *a.* Review excavation limits, construction access patterns, the site utilization program (see Section 5.4), and all other relevant temporary and permanent site constructions and movements to identify methods of establishing survey points that won't be destroyed by or during construction operations. These considerations will determine things such as:
      - Setback distances for excavation stakes
      - Sequence of surveys (if, for example, certain areas of the site must be established to allow room to proceed with other areas)
      - Quantities of survey points needed for the respective Subcontractors and trade Contractors to properly layout the remainder of the work
4. Prior to beginning any actual construction layout, the survey firm should begin with the verification of basic site information. This verification can be considered as really a portion of the preconstruction survey of Section 5.11, and should include as a minimum the verification of:
   *a.* Property boundary lines
   *b.* The Owner's correct establishment of the baselines and benchmark
5. The survey crew will then proceed to establish all control lines as discussed in step 3. As part of this initial survey effort, arrange for the survey crew to transfer the benchmark elevations of as many areas around the site as possible onto stable existing land features or constructions. Remember, once the building begins to go up, the original points will be obstructed. Be sure to have baseline elevations at every convenient location around the site.
6. The building footings and major site components will be laid out by the respective trades responsible for those portions of the work.
   *a.* The actual layout work will be done by the responsible trades.
   *b.* The Company will initially and periodically confirm that the procedure being used by the subtrades is appropriate and is being performed correctly.
   *c.* The actual layouts themselves will be checked as often as feasible by Company site personnel.
7. At the start of construction, after the initial site effort, verify alignment with the established building lines and benchmark elevations.
8. After footing placement and before foundation-wall construction, arrange for the return of the survey crew to pin the foundation and thereby physically

establish the actual major foundation-wall locations. The remainder of the foundation walls can be laid out from those points.

9. With the exception of only the simplest and smallest layouts, arrange for the survey crew to verify, at appropriate points of construction:
   - Anchor bolt locations
   - Bearing plate elevations
   - Beam pocket settings
   - Additional concrete structures
   - Slab elevations (spot-check several points to confirm specified flatness)
10. Throughout construction, Company personnel should periodically (often) verify for themselves:
    - Foundation centerlines and locations
    - Square points
    - Anchor bolts, bearing plates, and beam pockets
    - Plumbness and squareness of structural steel and all concrete structures
    - Alignment of beams, walls, columns, etc.—visually and with survey equipment
    - Slab elevations and flatness
    - Apparent relationship of actual contiguous construction to that anticipated from the documents
11. At the completion of the foundation, arrange for a certified survey of the building location. This information is now first needed to confirm error-free building location. From there it will be used to:
    - Provide as-built location confirmation (see Section 5.20) necessary for the Certificate of Occupancy and for the as-built requirement itself.
    - Establish the first portion of what will become the completed certified survey of the building when the remaining structure is completed, if one is deemed necessary. Review this requirement with the Project Manager.
12. At the completion of the building structure itself, including all major components (overhangs, towers, etc.), it may become necessary to provide a certified survey of the entire building. In some cases the final height of the building must be included in this certification. Review the requirement with the Project Manager and ensure that arrangements are made to provide all required information.
13. As a general note, the project characteristics themselves will greatly determine the amount of professional survey effort advisable throughout the construction period. Considerations for project size, complexity, proximity to property lines and other structures, and so on, may indicate the necessity to have professional survey crews periodically (monthly or weekly) return to the site to perform and confirm the Superintendent's checks described in item 10. This level of attention must be determined on an overall basis by Company senior project management at the project's onset, and must be effectively supervised and managed by the Superintendent throughout construction.
14. All original survey file information is to be forwarded to the home office for record, with a copy retained at the jobsite file.

## 5.13 Excavations—Special Precautions

### 5.13.1 General

Having performed the preconstruction survey in all the detail of Section 5.11, thereby at least generally confirming the overall accuracy of the existing visible information, the probability remains that:

- The actual configurations of underground constructions (pipes, conduits, wires, ductbanks, etc.) are different than those shown.
- Even if "substantially" correct, the *precise* locations of these items will vary to greater or lesser degrees from information on the plan.
- There will be items below the ground (hopefully not a long-distance phone cable) that are not shown at all.

Before *any* excavation proceeds in any previously undisturbed area of the site, the Superintendent is to:

1. Contact "Call-Before-You-Dig" or any other one-call system available for securing services necessary to positively locate all underground utility and communication structures
2. Use the *Notification Checklist Prior to Proposed Excavation* in Section 5.13.3, first to ensure that all appropriate individuals and companies are notified of the planned excavation, and second to document the effort

### 5.13.2 "Call-Before-You-Dig"

In every metropolitan area and in most other locations throughout the country, one-call systems have been put into place and maintained by associations of utility companies in the respective geographic areas. Generally it will be an 800 number listed under the name "Call-Before-You-Dig."

This phone number should be available from every utility in the area and from most insurance companies. Find out what the number is, and write it on the Sample Notification Checklist Prior to Proposed Excavation prior to reproducing the job supply of the form.

### 5.13.3 Sample Notification Checklist Prior to Proposed Excavation

The Sample Notification Checklist Prior to Proposed Excavation that follows must be completed by the Superintendent prior to any excavation in any previously undisturbed area of the project:

1. All underground systems must be physically located in the field.
2. Every individual or entity possibly affected by the construction must be notified prior to the work proceeding.
3. The effort to perform this research and notification must be properly documented and filed in the correct locations.

The purpose of the checklist, then, is to:

1. Provide a vehicle to aid in the coordination of this important effort, and to help eliminate oversight
2. Permanently record the names and phone numbers of all authorized parties contacted, along with the dates they were contacted and their inaction or actions taken
3. Transmit all necessary information regarding the project, the work to be done, and the dates when the work will proceed to each operator of every underground system

## 5.13.3
## Sample Notification Checklist Prior to Proposed Excavation

Project: ______________________ #__________
Location: ______________________________

| OPERATOR | PHONE | PERSON'S NAME | TIME NOTIFIED | DATE NOTIFIED | SITE ARRIVAL | ACTION TAKEN |
|---|---|---|---|---|---|---|
| Gas - Local | | | | | | |
| Gas - Transmission | | | | | | |
| Electric | | | | | | |
| Telephone - Local | | | | | | |
| Telephone - Lg Dist. | | | | | | |
| State Maintenance | | | | | | |
| County Maintenance | | | | | | |
| Local Water Dept. | | | | | | |
| Local Sewer Dept. | | | | | | |
| Local Highway Dept | | | | | | |
| Local Police Dept. | | | | | | |
| Local Fire Dept. | | | | | | |

**"Call-Before-You-Dig" or other one-call system in effect:** ______________________

INFORMATION TO BE TRANSMITTED TO THE ABOVE OPERATORS:

Company Name: ______________________
Company Address: ______________________
______________________
Company Telephone: ______________________

Company Field Rep. ______________________
Co. Field Address ______________________
______________________
Co. Field Telephone ______________________

Project Owner: ______________________
Plans Available At: ______________________
______________________

Location of Work: ______________________
______________________
______________________

Dates of Excavation: ______________________
Purpose of Excavation:
______________________
______________________
______________________
______________________
______________________

Record of Conversations, Meetings, Notes, & Remarks:
______________________
______________________
______________________
______________________
______________________
______________________
______________________
______________________
______________________
______________________
______________________
______________________
______________________

Checklist/Information prepared by: ______________________

Complete the form in its entirety prior to beginning excavation in any new area. Forward the original to be filed at the home office, and retain a copy in the jobsite Site Notification File.

## 5.14 Cutting Structural Elements

### 5.14.1 General

If it should ever become necessary or desirable to cut even a small portion of any structural element, *no* action is to be taken by Company personnel and *no* authorization is to be given to any Subcontractor without first fulfilling the requirements of the *Sample Structural Modification Authorization Form* of Section 5.14.2.

Types of structural elements falling under this consideration can include:

- Steel beams, columns, joists, and deck systems
- Steel struts, braces, and supports of any kind
- Concrete grade beams, bond beams, beams, wall panels, columns, supports, and bracing
- Structural studs, composite wall constructions, self-supporting walls, cantilever designs, any kind of spandrel construction
- Any sizable penetration or modification to masonry
- Anything that is unusual in any way, or that you are unfamiliar with

### 5.14.2 Sample Structural Modification Authorization Form

Before any such modification work can be allowed to proceed, *written* authorization must be secured from all authorized entities, specifically allowing the particular modification. This is necessary for *all* modifications—from those as complex as the relocation of structural components to the ones as simple as a 1.5-inch pipe penetration through the 60-inch web of a wide-flange steel beam.

In complex situations it is likely that the issue is so serious that it has or will warrant a large amount of attention with appropriate elaborate documentation of the design change and likely change order. In these cases the documentation and authority should be well taken care of.

If, however, this should for any reason not be the case, or if you find yourself with a "simple field condition" that would otherwise be handled quickly, verbally, and without the full contract procedure, the *Sample Structural Modification Authorization Form* that follows must be used. It will:

1. Describe the proposed modification in complete detail
2. Secure the approving signatures of all appropriate authorities
3. Provide complete documentation of the modification and authority

When the form is used, be sure to:

1. Describe the modification in a way that cannot be misunderstood. Include a sketch with all dimensions either on the modification form itself if possible, or as an attached supplement.

## 5.14.2
## Sample Structural Modification Form

Project:________________________________________ #_________ Date:_________________

This form summarizes modifications necessary to accommodate job conditions in order to allow construction to proceed in affected areas.

1. Description of need and proposed changes:

2. Indicate in drawing below or on attached drawings to be references below all structural components and proposed modifications. Include all dimensions and descriptions of materials:

**Authorization and Approval:**
The work described above is not to proceed until this Structural Modification Form is executed below by all authorized individuals as indicated:

| | | |
|---|---|---|
| Owner:__________ | Architect:__________ | Engineer:__________ |
| By:__________ | By:__________ | By:__________ |
| Signature:__________ | Signature:__________ | Signature:__________ |
| Date:__________ | Date:__________ | Date:__________ |

2. Tape a photocopy of the Modification Authorization Form on the jobsite and as-built sets of plans as the permanent record of the modification. Do this immediately as the form has been completed and signed.
3. Have the form signed in duplicate. Forward an original to the home office file of the respective bid package, and retain one at the corresponding jobsite file.

## 5.15 Control of Materials Embedded in Concrete

### 5.15.1 General

On projects where the Company is placing its own concrete, the considerations of this section are crucial to the avoidance of costly errors and rework due to the oversight of items required to be embedded or cast into concrete.

In those instances where the concrete work on the project is being performed by a Subcontractor or trade Contractor, the responsibility for the respective coordination is with those Subcontractors and trade Contractors. It does remain, however, in the Company's best interest to take whatever steps are possible to collectively help to minimize the possibility of error. This is not just a service for those who are actually performing the work, but a realization that any success in this area translates directly into project success in terms of avoidance of arguments and, most importantly, avoidance of problems that delay work on the critical path.

The remainder of this section assumes concrete placement by the Company and coordination efforts to be directly assumed by your own operation.

### 5.15.2 Sample Concrete Placement Checklist and Sign-Off Form

Depending upon the actual jobsite staff, the Superintendent, Concrete Foreman, or Field Engineer will be charged with the responsibility to complete the Concrete Placement Checklist and Sign-Off Form prior to each placement of concrete:

1. Completing the form will aid in the coordination of your own work.
2. Requiring the signatures of the foremen or other authorized Subcontractors and trade Contractors will force everyone to take close looks and to double-check their own work. Subsequent problems will accordingly be minimized.
3. Securing the signatures of these authorized representatives will directly release the Company from later criticisms of "failure to coordinate" in the event related problems do arise.

Insist on the procedure's being instituted at the very start of the project in even the simplest footing placements. It:

- Makes everyone familiar with the procedure
- Clarifies how importantly we consider each subtrade's responsibility for the coordination of its own work
- Establishes a routine so that everyone will know what is expected of them, and will be prepared to deal with it in each case in stride.

## 5.15.2
## Sample Concrete Placement Checklist & Sign-Off Form

Project:______________________________ #__________ Date: ____________________

1. Concrete Contractor/Subcontractor______________________________

2. Concrete Supplier:______________________________

3. Design Mix:______________________________ Required PSI:____________

4 Quantity to be placed: ________________ CY

5. Description & location of placement:

______________________________
______________________________
______________________________
______________________________
______________________________

6. Special material requirements (reinforcing, color, topping, finish, etc.):

______________________________
______________________________
______________________________
______________________________

7. Significant observations, remarks, and notes:

______________________________
______________________________
______________________________
______________________________
______________________________

8. Coordination of Work:
The signature below of the authorized representative of the respective company performing the work certifies that the work complies with the current requirements of the contract documents in every respect, and has been thoroughly coordinated with the work of all other potentially affected trades; considering dimensional locations, clearances and tolerances, operation, anchoring methods, materials used, that all materials used in their configurations are as approved in accordance with the Contract, and all other coordination issues have been properly accommodated:

| TRADE | COMPANY | INDIVIDUAL | SIGNATURE | DATE |
|---|---|---|---|---|
| Concrete | | | | |
| Steel Reinforcing | | | | |
| Masonry | | | | |
| Structural Steel | | | | |
| Misc. Steel | | | | |
| Sitework | | | | |
| Plumbing | | | | |
| HVAC | | | | |
| Fire Protection | | | | |
| Electrical | | | | |
| Communications | | | | |
| Other:__________ | | | | |
| Other:__________ | | | | |
| Other:__________ | | | | |
| Project Superintendent | | | | |

## 5.16 Construction Photographs

### 5.16.1 Description and Requirements

The job photo record begins with the preconstruction photo survey as described in Section 5.11.2. Ongoing construction photographs are the subject of this section. They are divided into regular progress photos and those required in special situations. There should always be a camera on-site and ready. Whenever there is any situation involving a significant question or the potential for changes, problems, etc., taking a photo of the area before any further discussion—and certainly before the area becomes further disturbed—must become the habit of the field staff. Film will always be the least expensive but most powerful agent of every negotiation, resolution, and settlement effort.

Resist the temptation to use wide-angle lenses in all photographs. While they make it easier to get more into the picture, they distort shapes, relative sizes, and perspectives. Remember that the objective is to provide an accurate record. Use lenses with "normal" focal lengths, and your photos will remain truer representations of reality.

### 5.16.2 Regular Progress Photographs

Regular progress photographs may or may not be required by the Owner as part of the General Conditions. If they are, the number of views, size of prints, and other criteria may be specified. If progress photos are not an Owner requirement, arrange for regular progress photos anyway as a Company record. In this effort:

1. Set aside a regular day each month to conduct the photo record. Use the first Monday or first Wednesday, for example. This will help establish a routine that will be easy to monitor.
2. Use a 35-mm camera of good quality. Autofocus and autoflash systems are highly desirable, but not necessary. If at all possible, using a camera back that automatically dates the film is one of the greatest available features.
3. Generally, follow the guidelines for preconstruction photographs in Section 5.11.2. There should be no need for the level of detail of the preconstruction set. These regular photos are intended to show the state of progress, and not necessarily aimed at catching every problem. The latter is discussed in Section 5.16.3.
4. Consider supplementing the photo set with a regular video. As with the preconstruction effort, the video is not a substitute, but will be a thorough record of the entire project.
5. Identify and date each regular survey. Include a copy of the Daily Field Report that records the photo effort.
6. Have each photo set developed immediately and sent to the home office Progress Photo File.

### 5.16.3 Special Situations

"Before-During-After" photos are necessary throughout each situation involving actual or *potential:*

- Change orders
- Claims
- Backcharges
- Insurance claims
- The "surprise du jour"
- Any other special situation

In these kinds of events, fast action is necessary to gain the maximum advantage possible. A "precondition" photo or series of photos will be the best record to confirm the actual state of affairs prior to a changed work sequence.

If the duration of the anticipated sequence or activity is relatively long, progress photos of the specific sequence should be taken.

An "instant" camera may prove to be a valuable addition to the project photo effort. Again, it is to be used as a supplement and can only become a substitute for better-quality photos in the simplest of situations (a cleanup backcharge?).

The principal advantages of instant pictures is that they immediately display the specific information that has been recorded in the photo. It is therefore immediately evident if your photos indicate everything intended, or if additional photos, perhaps from other angles, will be necessary. This may be particularly important if conditions are likely to change quickly, leaving little time for conventional photos to be reviewed. Even so, one-hour developing is becoming readily available.

### 5.16.4 Use of Photograph Layout Form

Regular progress photographs for the records will not require any identification other than that described in this section and in Section 5.11.2 for the preconstruction photo survey.

Special photographs are, however, taken for a specific purpose and must be so identified. In the worst conditions of arbitration or litigation, each photo must be properly and completely identified along strict criteria. If it is not, even the best photo will not be allowed consideration. Even if the photo's uses will not become so extreme, references to special photos most often come at points in time long after the original incident. Memories fade, orientations get confused, and so on.

It is therefore important that each significant photo be properly identified if it is to be used to support a charge, contention, or any special circumstance.

Using the Sample Photograph Layout Form that follows will guarantee that each photo that will immediately or eventually be used in any presentation or as support for any kind of charge or contention will be correctly identified with:

1. Names of project and Owner, Company project number
2. Photo location, area identification
3. Photo orientation, direction of view
4. References to any appropriate correspondence, field reports, or anything else that ties it directly to the detailed project record

### 5.16.5 Sample Photograph Layout Form—Completed Example *(page 5.67)*

### 5.16.6 Sample Photograph Layout Form—Blank Form *(page 5.68)*

## 5.17 Managing Time and Material

### 5.17.1 General

Time and Material (T&M) considerations apply in the performance of:

1. Company work to be charged directly to the Owner
2. Subcontractor work to be charged to the Company that will be passed through as an Owner change order
3. Subcontractor work to be charged to the Company that will not be passed through as an Owner change order

The procedure will be followed in all instances where:

1. The work in question is an agreed extra in any of the preceding categories
2. The work in question either is in dispute as an extra or is still subject to further review and/or reconsideration

In cases of a Company charge to the Owner, T&M is a final option normally included in the Change Clause of the general contract. (Refer to Section 3.3.7 for a complete description.) It should be pursued immediately as an approved pricing form in any instance where lump-sum or unit-price methods fail to secure agreement on costs.

For the company performing the T&M work, the advantages include:

1. Little or no risk of cost overruns. Everything is billed at cost-plus. Coordination and efficiency problems are "absorbed" by the work.
2. The work can proceed with less delay than that associated with change order approval.

For the Owner or Company who must pay for the charges, the disadvantages include:

1. Assumption of nearly all risk of coordination effectiveness and performance efficiency, under conditions where:
   *a.* Incentives to do these things well are effectively eliminated. With little risk of failure, there is reduced incentive for adequate attention.
   *b.* There is a financial conflict of interest. The more efficient the working force is, the less will be the final change order value.
2. Complete open-ended nature of both final cost and time.

These are *significant* disadvantages that are not readily overcome.

## 5.16.5
## Sample Photograph Layout Form (Completed Example)

Project: FIREHOUSE ADDITION # 9424 Date: JUNE 16, 1994
Taken By: MARK LEONARDO Time: 9:20 (AM) PM

Location: NORTH END OF CORRIDOR 224 - CRACK AT DUCT PENETRATION
Orientation (Indicate on Key Plan): VIEWING NORTH
Remarks: PHOTO #1 - FROM SOUTH END OF CORRIDOR; SHOWS SURROUNDING CONST.
PHOTO #2 - FROM CORRIDOR MID-POINT; CLOSE-UP OF WALL SECTION

Place photos /Assign numbers to multiple photos / Attach this form to large photos.

(Photo) ①

(Photo) ②

Key Plan of Photo Area(s):

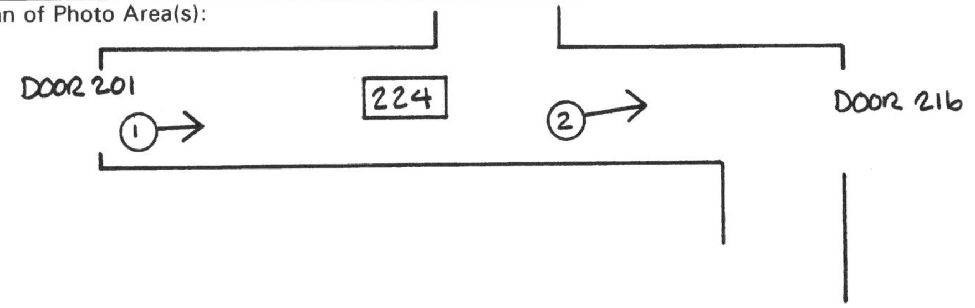

DOOR 201
224
①→
②→
DOOR 216

## 5.16.6
## Sample Photograph Layout Form

Project: ______________________________ #________ Date: ________________
Taken By: ______________________________________ Time: __________ AM PM

Location: ________________________________________________________________
Orientation (Indicate on Key Plan): _________________________________________
Remarks: ________________________________________________________________
__________________________________________________________________________
__________________________________________________________________________

Place photos /Assign numbers to multiple photos / Attach this form to large photos.

__________________________________________________________________________

Key Plan of Photo Area(s):

### 5.17.2 Field Staff and Company Responsibility

Responsibility of the field staff and of the Company itself with respect to T&M is therefore divided between:

- Controlling T&M work by Subcontractors on behalf of the Company
- Effectively managing T&M work to be charged to the Owner in a responsible, equitable manner

**Subcontractor T&M work.** On balance, it is a *very* rare situation that truly justifies *any* T&M work by any Subcontractor to the Company. As a Company policy, Subcontractor T&M should only be considered in those instances where for very real reasons complete pricing and time estimates are truly not possible or advisable.

If you find yourself considering the approval of any T&M work, take another hard look. Be aware that you will be expected to explain to senior management why the pricing could not be secured properly before any changed and/or additional work on the part of a Subcontractor was authorized to proceed. Realize at the start that:

1. If you restructure the proposed T&M work to a lump-sum arrangement, you:
   *a.* Return the risk of success, along with all coordination responsibilities and performance incentives, back to the Subcontractor
   *b.* Free yourself from the extreme responsibilities and daily efforts necessary for management and verification of the work
2. If you allow proposed T&M work to proceed, you and the Company directly assume these substantial performance issues.

In any event, the Superintendent is not authorized to allow any T&M work without express approval by the Project Manager.

**T&M to be passed through to the Owner.** In situations where there is any delay in the resolution of a change order's final cost, or if there is any real reason why complete pricing is not feasible, T&M may turn out to be the option of choice for the Company. The principal reason would be the ability to proceed with the work, thereby mitigating the net effect on the project.

In every approved T&M situation, whether the work is being done by Subcontractors and passed through to the Owner, or the work is being done by Company forces and charged directly, there is a primary obligation on the part of the entire field staff and of the Company to shoulder our own responsibilities completely. In this regard, any T&M work will be approached as if it were lump-sum:

1. With the same considerations and attention given to coordination, management efficiency, and every cost-control effort
2. Performed with the overall idea of performing a service for the Owner

### 5.17.3 T&M Procedure

In all cases where T&M work has been authorized to proceed, the following procedures and precautions can be taken to improve control:

1. *Have effective standard subcontract language.* Be sure that your subcontracts contain a standard clause disallowing any T&M work unless:

   *a.* It has been approved in writing in advance
   *b.* T&M tickets are signed *daily*; otherwise the T&M work expressly will not be recognized

2. *Daily means daily.* A 2-inch stack of T&M tickets presented near the end of the altered work is impossible to decipher with any degree of effectiveness.

   *a.* If they had been prepared as a group, you can guarantee that each ticket will likely be "generous" in its assignment of man-hours to extra (as opposed to contract) work.
   *b.* Even if you've got your own detailed records with which to check, the process will be long, tedious, and error-prone.

   In contrast, strictly enforce the requirement for *daily* review. Verify all information on a current basis. Always take the presenter to task in explaining each day why any portion of the work represented to be extra is not actually part of its contract.

   Emphatically emphasize the fact that T&M tickets that have not been signed off *on the day that the work was performed* will not be recognized as extra work. Consider using the *Sample Letter to Subcontractors Regarding T&M Submission Requirements* in Section 5.17.4 for help in placing emphasis on the requirement.
3. *Verify actual hours.* If multiple items are being billed for on a T&M basis, or if the same individuals are being used concurrently for contract work, check the total number of hours billed on a given day. Do they add up to twelve hours for an eight-hour day? If so, such abuses must be detected early. You will have every right to be offended and to let everyone know that advantage is being taken. Get control over the situation at the start.
4. *Verify labor classifications.* Are you getting billed for a 50 percent apprentice at a journeyman's rate? Is a high wage rate from another local being added to the stack of forms? Are prevailing wages truly being paid to the workers?
5. *Verify overhead and profit application.* Are overhead and profit rates being billed at the subcontract (general contract) change order rates, or are they being creative? If you have agreed to pick up the cost of premium time charges, is overhead and profit being applied to the premium-time portion of the bill without your authorization?
6. *Take photographs as appropriate.* Film is cheap. In every situation involving *any* question, immediately take photographs. Refer to Section 5.11, *Preconstruction Survey,* and Section 5.16, *Construction Photographs,* for more in this regard.
7. *Evaluate production rates.* As soon as possible, and at points throughout the T&M work, review rates of progress and production efficiency and compare to those you can reasonably expect.

   *a.* Even the smallest liberties being taken at the start must be put back into perspective immediately.
   *b.* Significant and/or chronic abuses must be dealt with decisively.

   Tolerate *no* abuse. Get help from the home office if necessary to put appropriate pressure on the offending party. Consider:

   - Arranging for Company senior management to resolve the problem with the offender's senior management
   - Rescinding the authorization to proceed

- Notifying the party that certain portions of the work will not be recognized for payment

  In so doing, be prepared with specific criticisms backed up with hard documentation. Be able to definitively support every contention.

When presenting Company T&M tickets to the Owner, it is important to understand that the Owner has the same concerns toward the Company as you have toward Subcontractors. Owners, however, can experience even greater feelings of loss of control. From their perspective:

- The work is difficult or impossible to check
- The work needs to be watched with a microscope
- The totals *always* seem to add up to more than expected (or hoped for)
- You are operating with little or no risk, and therefore without any motivation for cost control or production efficiency

The results of all this can range from simple bad feelings over the situation to major disputes over T&M components.

In order to counter these issues, police your own work in the same way as was recommended earlier to verify the validity of work by others. Involve the Owner throughout the T&M process. Demonstrate your attention, awareness of the issues, and concern for equity. If this is all done with genuine concern for maintaining equity in the situation, there won't be major problems later.

A final consideration in T&M work refers to those instances where the responsibility to perform certain work is not clear or is the subject of dispute. If in such circumstances the Company has determined that in the interest of the project the work must proceed either pending a final decision or while anticipating a later claim, use the T&M procedure and tickets to keep accurate, indisputable records of the actual work performed. Refer to Section 5.17.6 for specific direction in this regard.

### 5.17.4 Sample Letter to Subcontractors Regarding T&M Submission Requirements *(page 5.72)*

The Sample Letter to Subcontractors Regarding T&M Submission Requirements that follows is designed to follow through with the recommendations of item 2 of the procedure described in order to support the effort to police each T&M Subcontractor's requirement to submit the tickets in proper, complete form and on time. Specifically, it notifies the respective Subcontractor that:

1. *Daily* authorization and acknowledgment of work performed is absolutely the requirement.
2. T&M tickets not approved on the day that the work was performed *will not be recognized* as additional cost.
3. Labor and material information requires detailed breakdowns and complete substantiation of all costs.
4. Signatures of field personnel confirm only the fact that certain work was actually performed with certain forces on that day. This does not acknowledge or

## 5.17.4
## Sample Letter to Subcontractors
## Regarding T & M Submission Requirements

| Letterhead |
|---|

(Date)

To: (Subcontractor)

RE: (Project)
(Company Project #)

SUBJ: (Company Change File No.)
(Change Description)

Mr. (Ms) :

On (Date), you were directed to proceed with the subject work on a Time & Material basis in accordance with the provisions of (insert the appropriate general or subcontract reference). Conditions of the arrangement are confirmed as follows:

1. All T&M tickets are to be signed daily by authorized personnel. Tickets not so approved on the day that the work was actually performed will not be recognized as an additional expense.

2. Precise labor classifications are to be described on each ticket.

3. Material invoices are to be included.

4. The signatures of field personnel confirm the fact that certain work was performed by certain forces on that day. They do not acknowledge or agree that the work itself is in addition to your contract, or that the rates charged are accepted without further review and confirmation.

Thank you for your cooperation.

Very truly yours,

COMPANY

Project Manager

cc: Jobsite
File: Vendor File: ________
Change File ( )
CF

agree in any way that the work itself is extra or that the rates charged are accepted. These require agreement at another level.

The real purpose of the letter is to set the tone for the type of attention that will be given the T&M work. If the proper routine is established from the start, the entire process will move ahead much more efficiently and equitably.

### 5.17.5 Sample T&M Form (Daily Report of Extra Work) *(page 5.74)*

The Sample T&M Form (Daily Report of Extra Work) that follows is to be used to fulfill the requirements of this section. A supply should be given to each Subcontractor about to work under any T&M arrangement in order to create uniformity in T&M reporting and to help make their reviews more routine.

No matter which forms are finally used for any work, be certain that each one is prepared with *all* relevant information and secured *daily*.

### 5.17.6 Using T&M Records to Support Changes and Claims

An important use of the T&M procedure occurs in those instances where the responsibility for certain work as part of the Owner/Contractor agreement is not called for, is not clearly extra, or is clearly extra but is, at least for the moment, in dispute.

If circumstances lead project management to decide to proceed with the disputed or questioned work, either pending a final decision or in anticipation of a later claim, use the T&M procedure and forms to keep accurate, indisputable records. In order to do this:

1. Prepare the T&M forms as you would if the Owner were clearly paying for the work.
2. Arrange with the Owner's site representative to have the daily forms signed only to acknowledge the fact that certain work was performed by certain workers on that day.
   *a.* Clarify that it is the documentation of fact only, and that you are aware that the Owner has not agreed to any payments for the work at this time.
   *b.* If necessary, use the Sample Letter to Owner Regarding Acknowledgment of Actual Work Performed of Section 5.17.7 to help accomplish this.
3. If for any reason the Owner representative refuses to even provide this basic acknowledgment, present the forms to him or her *on a daily basis* anyway. Upon each refusal, write the precise circumstances on the respective T&M form, including the fact that it was presented for acknowledgement, or that the work itself was reviewed by the representative who refused to sign.

This effort will become extremely valuable in later Company efforts to settle the issue. The problem will be narrowed to the issue itself, with little time and energy necessary to substantiate the costs.

## 5.17.5
## Sample T&M Form
## (Daily Report of Extra Work)

### DAILY REPORT OF EXTRA WORK

Date: ____________________
Project: ____________________ #: ________
Charged To: ____________________

Contract Work: _____
Change Order: _____
Work Done Under Protest: _____ YES _____ NO
Work Part of a Claim: _____ YES _____ NO

Bid Package: ____________________
Owner CO #: ____________________
Company CO File #: ____________________

Description of Work:

**(A) LABOR**

| Class | Quan | Hrs | Rate* | Amount |
|---|---|---|---|---|
| | | | | |
| | | | | |
| | | | | |
| | | | | |
| | | | | |
| | | | | |
| | | | | |
| | | | | |
| | | | | |

**Total Labor (A)** $________

* Includes applicable health, welfare, pension, insurance, and taxes.

**(B) MATERIAL**

| Description | Quan | Unit Price | Amount |
|---|---|---|---|
| | | | |
| | | | |
| | | | |
| | | | |
| | | | |
| | | | |
| | | | |
| | | | |
| | | | |

Subtotal $________

Sales Tax $________

**Total Material (B)** $________

**(C) EQUIPMENT**

| Size/Class | Quan | Hrs | Rate | Amount |
|---|---|---|---|---|
| | | | | |
| | | | | |
| | | | | |
| | | | | |
| | | | | |
| | | | | |
| | | | | |
| | | | | |
| | | | | |

**Total Equipment (C)** $________

Prepared By: ____________________

Approved By: ____________________

**GRAND TOTAL (A+B+C):** $________

### 5.17.7 Sample Letter to Owner Regarding Acknowledgment of Actual Work Performed *(page 5.76)*

The Sample Letter to Owner Regarding Acknowledgment of Actual Work Performed that follows is an example of the confirmation that follows through on the recommendations of the preceding section. It is used to advise the Owner that you will expect daily verification of actual work performed for any item of work that is in dispute for any reason. It is to be used in every instance where:

- The decision for the responsibility of the work is pending.
- You have been directed to proceed with work that is definitely or may be extra to the contract.
- Project management has decided that it is in the best interest of the project to proceed with the work in question, rather than to stop until the issue is completely resolved.

The letter advises the Owner that:

1. You are proceeding under protest per the Owner's direction, and do not agree to be responsible for the extra costs
2. You will be preparing T&M forms on a daily basis for the purpose of confirming actual work performed
3. You expect the Owner's representative at the site to sign the daily tickets; specifically to acknowledge the accuracy of the reported information regarding labor used and resources consumed
4. You acknowledge that the Owner's signature in no way agrees to the idea of an extra to the contract at this time

## 5.18 Field Purchases Procedure

### 15.8.1 General

The field purchases procedure is to be followed in all instances where minor purchases are made directly by jobsite personnel from locations near the field. Only Company employees are to make any purchases on behalf of the Company.

The purposes of the procedure are to:

1. Confirm that the item(s) considered for purchase are not already available from the home office or another jobsite
2. Control the amount of materials purchased and buy only those actually needed for the work item in question
3. Limit "supplementary" or otherwise inappropriate items (such as small tools, blades, bits, and anything else that is the responsibility of someone else)
4. Ensure that the materials purchased are properly identified and clearly assigned to those ultimately responsible for the cost (change orders, backcharges)

## 5.17.7
## Sample Letter to Owner
## Regarding Acknowledgement of Actual Work Performed

Letterhead

(Date)

To: (Owner)

RE: (Project)
(Company Project #)

SUBJ: (Company Change File No.)
(Change Description)
Acknowledgement of Actual Work Performed

Mr. (Ms) :

On (Date), you directed the subject work to proceed without agreeing to a change order.

Please be advised that we are proceeding with the subject work under protest, strictly in the interest of minimizing impact on the project.

We will be preparing daily Time & Material tickets to document the actual work performed, along with all resources used. They will be presented each day to your on-site representative for signature. We recognize that this signature will acknowledge only the facts of the information contained in the respective T&M tickets, and does not at this time indicate acceptance by you of the responsibility for the work.

Very truly yours,

COMPANY

Project Manager

cc: Jobsite
Owner Field Representative
Architect
File: Change File ( )
CF

## 5.18.2 Procedure

The Project Engineer, the Superintendent, and the individual actually making the purchase all have responsibilities to fulfill in the procedure:

1. Project Engineer

   *a.* The *Field Purchase Order Log* is established for the individual project and placed in the Telecon Log Book (see Section 5.18.3).
   *b.* The Project Engineer or designate will be responsible to assign field purchase order numbers at the time the appropriate Company on-site individual calls for authorization.
   *c.* Prior arrangements from the Site Superintendent must be made before any field purchase order number is given. At that time, the following information must be entered into the Field Purchase Order Log:

      (1) Description of purchase
      (2) Where the purchase will be made
      (3) Reason for purchase, including adequate description of cost responsibility and reference to appropriate change order or backcharge file.

   *d.* The field purchase order number is assigned to the purchase.
   *e.* The *Field Purchase Order Form* is then completed (see Section 5.18.4). Information provided on the form includes:

      (1) Reason for purchase
      (2) Complete change order or backcharge description, file reference, and all other appropriate information and references

   *f.* A copy of the Field Purchase Order Form is distributed to:

      (1) Accounts Payable
      (2) Any referenced change order file
      (3) Any referenced backcharge file
      (4) Any affected subcontractor or supplier file

2. Superintendent and field personnel

   *a.* Before leaving the site to purchase anything, the purchase must be set up with the Project Engineer or designate. Include information as required in 1*c* and get a field purchase order number.
   *b.* While at the purchase site, a call from the Vendor is required to confirm the field purchase order number. At that time the complete description of the items purchased and the total cost of the purchase must be available.

3. Accounts Payable

   *a.* After the Field Purchase Order Form has been received from the Project Engineer, it is to be placed into the Payable File for that project.
   *b.* When the respective invoices are received from the Vendors, the Field Purchase Order Form is attached to them. The invoice now has the complete detailed information included with it.

## 5.18.3 Sample Field Purchase Order Log *(page 5.78)*

## 5.18.4 Sample Field Purchase Order Form (page 5.79)

## 5.18.3

## Sample Field Purchase Order Log

### PURCHASE ORDER LOG

Project: ____________________ No: ____________

| P.O # | AMOUNT | APPR BY | DATE | FIRM | DESCRIPTION | BACK CHARGE | CO FILE # |
|---|---|---|---|---|---|---|---|
| | | | | | | | |
| | | | | | | | |
| | | | | | | | |
| | | | | | | | |
| | | | | | | | |
| | | | | | | | |
| | | | | | | | |
| | | | | | | | |
| | | | | | | | |
| | | | | | | | |
| | | | | | | | |
| | | | | | | | |
| | | | | | | | |
| | | | | | | | |
| | | | | | | | |
| | | | | | | | |
| | | | | | | | |
| | | | | | | | |
| | | | | | | | |
| | | | | | | | |
| | | | | | | | |
| | | | | | | | |
| | | | | | | | |
| | | | | | | | |
| | | | | | | | |
| | | | | | | | |
| | | | | | | | |
| | | | | | | | |
| | | | | | | | |
| | | | | | | | |
| | | | | | | | |
| | | | | | | | |
| | | | | | | | |
| | | | | | | | |
| | | | | | | | |
| | | | | | | | |
| | | | | | | | |
| | | | | | | | |
| | | | | | | | |

## 5.18.4

## Sample Field Purchase Order Form

# FIELD PURCHASE ORDER

TO:

PROJECT: ______________________ ORDER NO.: ______

PROJECT NO.: ______________________

ORDER DATE: ______

Confirmation of verbal order: ____ Ship Via: ____________

F.O.B. Point: ____________

Tax Exempt: ____YES ____ NO Tax Exempt No: ____________

DESCRIPTION OF NEED:

Bid Package Responsibility: ____________ Change Order #: ____________ Company File #: ____________

| Item # | Description | Quantity | Unit Price | TOTAL |
| --- | --- | --- | --- | --- |
| | | | | |
| | | | | |
| | | | | |
| | | | | |
| | | | | |
| | | | | |
| | | | | |
| | | | | |
| | | | | |

Sub Total: $ ____________

Sales Tax: $ ____________

Authorized By: ______________________ TOTAL: $ ____________

## 5.19 Winter Precautions

### 5.19.1 General

Winter precautions for projects in freezing climates are too often gone about on an uncoordinated, as-problems-become-apparent basis. The purpose of this section is to identify the issues and provide a straightforward method to organize a coordinated approach to this potentially expensive jobsite condition.

The considerations for winter precautions generally boil down to:

1. The physical constructions and services needed to properly protect the work, and to maintain such precautions
2. Determination of the complete responsibility to pay for the work, fuel, and materials involved

### 5.19.2 Subcontractor, General Contractor, and Owner Responsibilities

It is rare that the complete responsibility for all winter precautions lies only with the General Contractor. It is important to be aware from the start that the responsibility to provide any winter precautions at all is specifically related to the situation that can be anticipated from the bid documents, particularly with respect to the originally anticipated project duration at the time the contract was executed. This is the only circumstance under which the immediately following remarks apply.

Where it has clearly been anticipated as part of the original project cost and time, the General Contractor will generally provide those precautions and maintenance items necessary, as associated with the overall protection of the facility itself, including items such as:

- Closing in of open floors
- Providing space heaters
- Maintaining fuel supply to space heaters

Each respective Subcontractor will generally be responsible for *all* specific precautions and protections associated with any particular item of work or piece of equipment. Refer to the Pass-Through Clause of Section 3.3.9 for related discussion.

In any instance of delay, interference with, and/or disruption of planned sequences, the entire responsibility picture is likely to change:

- A Subcontractor's delay in delivering permanent heating equipment or in securing system operation may force the need for unanticipated temporary heating equipment.
- Any Subcontractor's delay in the work of the critical path may place portions of the work into freezing conditions that would otherwise have been done without the need for such protection by the General Contractor or another Subcontractor.

**Example:** The Concrete Subcontractor delays the foundation for exterior architectural masonry, causing the need to provide protection and heat for the masonry.

- The Owner (or design professionals) delay the work in any way (change orders, lack of timely action, etc.), placing work into freezing conditions.

These types of circumstances alter dramatically the responsibility to pay for establishing, maintaining, and removing winter precautions. The complete responsibility, then, for each anticipated item must be confirmed with the Project Manager prior to proceeding with any such work as part of any direct Company expense.

### 5.19.3 Winter Precautions Checklist

*(page 5.82)*

The Winter Precautions Checklist that follows is to be used as an aid in:

- Reviewing job conditions
- Assessing responsibilities for specific precautions
- Determining adequate precautions necessary for specific areas of work
- Confirming that appropriate arrangements have been made to provide for each required precaution

Use the checklist:

1. First as a meeting agenda between the Superintendent and the Project Manager to confirm all conditions
2. Second as a Subcontractor meeting agenda to confirm all that will be done by the respective trades, including timetables

File the completed checklist in:

- Home and field office files for winter precautions
- All related change order files
- Any related backcharge files
- Every affected Subcontractor or trade Contractor file
- The Correspondence File (CF)

## 5.20 As-Built Drawings

### 5.20.1 General

As-Built Drawings are required by nearly every specification for projects of any size. In those rare conditions where the As-Built Drawings are not specifically called for, they will be provided anyway as a Company requirement.

Their purpose is to serve as a permanent record for the Owner regarding all actual conditions relative to those originally designed, to note dimensional deviations not documented anywhere else, and to consolidate the *identification* of the modifications that have occurred throughout the construction period. The information is used to aid in future design, construction, and maintenance. As-Built Drawings are *not* there to repeat the detailed information of any change that is properly documented in the respective files.

# 5.19.3 Winter Precautions Checklist

YES NO

**A. GENERAL PROJECT STATUS**

As of ______________________

1. Building portions satisfactorily closed to Weather:
   a. Roofs & Flashings ___ ___
   b. Doors & Windows ___ ___
   c. Building Skin ___ ___
   d. ______________________ ___ ___
   e. ______________________ ___ ___
2. Permanent heating system usable for temporary heat:
   a. Electrical ___ ___
   b. HVAC ___ ___
3. Interior pipes/systems subject to freezing:
   a. Remarks: ______________________
4. Permanent Source of Power Available: ___ ___
5. Temporary Power Necessary: ___ ___
   a. Remarks: ______________________
6. Permanent source of fuel available: ___ ___
7. Temporary fuel necessary: ___ ___
   a. Remarks: ______________________

**B. CONTRACT ASSESSMENT**

1. Temporary heat required between (dates): ____________ and ____________
2. Responsibility to provide temporary heat:
   a. Owner ___ ___
   b. Prime Contractor or Const. Manager ___ ___
   c. Sub or Trade Contractor(s)
      1) ______________________ ___ ___
      2) ______________________ ___ ___
      3) ______________________ ___ ___
3. Responsibility to provide temp. protect.
   a. Owner ___ ___
   b. Prime Contractor or Const. Manager ___ ___
   c. Sub or Trade Contractor(s)
      1) ______________________ ___ ___
      2) ______________________ ___ ___
      3) ______________________ ___ ___
4. Temporary heat/protection now required because of delay: ___ ___
5. If (4) yes, who is responsible:
   a. Owner ___ ___
   b. Prime Contractor or Const. Manager ___ ___
   c. Sub or Trade Contractor(s)
      1) ______________________ ___ ___
      2) ______________________ ___ ___
      3) ______________________ ___ ___
   d. Reasons/Remarks: ______________________
6. If (5) is Owner:
   a. Change Order File established ___ ___
   b. C.O. acknowledged by the Owner ___ ___
   c. Is a claim necessary (denied C.O.) ___ ___
   d. If (6.c.) yes:
      1) Written notification made: ___ ___
         Date: ______________________
         To: ______________________
      2) Documentation provided: ___ ___
7. If (5) is Subcontractor or Trade Contractor:
   a. Has backcharge procedure begun ___ ___
   b. Written Backcharge Notice sent ___ ___
   c. Responsibility accepted ___ ___
8. Estimated cost of temporary services (Attach detailed estimate forms)
   a. Protection $__________
   b. Heating Equip. $__________
   c. Heating Fuel $__________
   d. Light & Power $__________
   e. Total $__________

**C. OVERALL JOB PRECAUTIONS**

1. Arrangements made to secure:
   a. Temporary protection materials ___ ___
   b. Temporary enclosure materials ___ ___
   c. Continuous fuel supply ___ ___
2. Temporary heating equipment is:
   a. Of adequate size & type ___ ___
   b. Is maintained / fully operational ___ ___
   c. Of type(s) allowed by codes ___ ___
   d. Situated in safe manner relative to pedestrians, traffic, building materials, and ventilation ___ ___
   e. On a service/maintenance schedule ___ ___
3. Temporary fuel is:
   a. On hand and in adequate supply ___ ___
   b. Properly and safely stored ___ ___
   c. On a set refueling schedule ___ ___
4. All water pockets have been eliminated:
   a. Roof areas ___ ___
   b. Pavement and graded areas ___ ___
   c. Sleeves, inserts, chases & openings ___ ___
   d. Other: ______________________ ___ ___
5. Arrangements have been made for:
   a. Snow plowing/removal ___ ___
   b. Equipment cold weather protection ___ ___
   c. Vehicle Maintenance ___ ___
6. Precautions taken to protect exposed work:
   a. Exposed piping protected, drained, or heat traced ___ ___
   b. Recently placed work (concrete, formwork, reinf. steel, masonry, etc.) ___ ___

| | YES | NO |
|---|---|---|
| 7. All project areas have been adequately marked to avoid damage during snow removal: | | |
| a. Parking areas | ___ | ___ |
| b. Entrances, exits, gates, passageways | ___ | ___ |
| c. Pedestrian traffic areas | ___ | ___ |
| d. Material and fuel storage areas | ___ | ___ |
| 8. Any necessary photographs of all pre-winter jobsite conditions taken for record | ___ | ___ |

**D. SPECIFIC WINTER PRECAUTIONS**

1. Item of Work: ______________________________

Location: ______________________________

Party Responsible: ______________________________
Specific Precautions taken: ______________________________

Precaution start date: ______________________________
Anticipated end date: ______________________________
Remarks: ______________________________

2. Item of Work: ______________________________

Location: ______________________________

Party Responsible: ______________________________
Specific Precautions taken: ______________________________

Precaution start date: ______________________________
Anticipated end date: ______________________________
Remarks: ______________________________

3. Item of Work: ______________________________

Location: ______________________________

Party Responsible: ______________________________
Specific Precautions taken: ______________________________

Precaution start date: ______________________________
Anticipated end date: ______________________________
Remarks: ______________________________

4. Item of Work: ______________________________

Location: ______________________________

Party Responsible: ______________________________
Specific Precautions taken: ______________________________

Precaution start date: ______________________________
Anticipated end date: ______________________________
Remarks: ______________________________

5. Item of Work: ______________________________

Location: ______________________________

Party Responsible: ______________________________
Specific Precautions taken: ______________________________

Precaution start date: ______________________________
Anticipated end date: ______________________________
Remarks: ______________________________

6. Item of Work: ______________________________

Location: ______________________________

Party Responsible: ______________________________
Specific Precautions taken: ______________________________

Precaution start date: ______________________________
Anticipated end date: ______________________________
Remarks: ______________________________

### 5.20.2 Procedure

1. Immediately at the start of the project, one complete set of plans, specifications, and addenda is to be sent to the jobsite clearly marked as "As-Builts." There is no need to "post" the addenda; just include it as part of the set. These documents are *not* to be used for construction. They are to be properly filed and kept in good condition.
2. Review the contract documents to determine any specific conditions required by the Owner for preparation, maintenance, and delivery of the As-Built Drawings. Comply in every respect.
3. It is the responsibility of the Project Engineer, Site Superintendent, and any other field staff to verify that any deviation between actual construction and that as originally designed is in fact properly authorized and documented as such prior to allowing such deviating work to proceed. These will include items such as:

   - Approved change orders
   - "Clarifications" not involving cost or time
   - Accommodations of field conditions that are slightly different than those originally anticipated
   - Actual locations and configurations of existing underground lines and construction as they are uncovered during the course of the work
   - Actual locations of new underground work if at all different from the plan locations

4. The Superintendent is to record in sufficient detail all dimensional deviations and all references to the detailed change records on *both* the jobsite document set and the As-Built Drawings *as they occur*. All such additions, deletions, or changes are to:

   *a.* Be indicated in red pencil or pen
   *b.* Be dated
   *c.* Include *clear* reference to appropriate authority for the modification, such as:

      - Change order file number
      - Job meeting item number
      - Conversation and confirming memo with name, conditions, etc.
      - Structural Modification Authorization Forms (see Section 5.14.2).

5. In the case of change orders and detailed clarifications, it is not necessary to redraft the detail of the change. Cloud the area affected by the respective change or clarification, and reference the appropriate change order number or other complete reference.
6. Whenever possible, tape a photocopy of any "SK" or other available sketches on the contract set. It is most likely that the only room will be on the back of the previous page. In this case, simply note "Taped Opposite" on the modified plan. Include copies of any Structural Modification Authorization Forms (see Section 5.14.2).
7. It is the Project Engineer's and the Superintendent's responsibility to police each major Subcontractor or trade Contractor to include their own as-built information on the Company field set and As-Built Drawings *on a weekly basis*. This information should be confirmed monthly by the Project Manager as an express condition of payment. These Contractors include at a minimum:

   - Concrete
   - Structural steel

- Plumbing
- HVAC
- Fire protection
- Electrical
- Controls
- Communications

8. Confirm final as-built configurations required by the contract prior to delivery to the Owner. It may, for example, be required to transfer the information to a set of mylars provided by the Owner. In such cases it is not necessary to transfer any supplemental documents taped to the plans as discussed in item 6. The references will be adequate.
9. Include all engineered layouts, confirmations, and certifications provided (see Section 5.12.3) and all certified As-Built Drawings by all trades required to provide them (fire protection, for example).
10. Hand deliver completed as-built documents to the Owner, and have them signed for by an individual authorized to receive them.

Section

# 6

# Safety and Loss Control

## 6.1 Section Description and Company Policy

### 6.1.1 General

Safety responsibility begins with a moral obligation. The jobsite simply must not pose any unnecessary current or future risk to any individual working in it or passing through it.

Furthermore there are real considerations for production efficiency and cost control as related to minimized lost time, and the legal issues following incidents of exposure, injury, and death. As it turns out, it is profitable for every company and individual, both financially and in many other ways, to give safety the full attention that it deserves.

This section of the Manual describes the safety policy of the organization, the responsibilities of each company and individual on the jobsite with respect to safety, the components of the ongoing safety program, and specific procedures to properly accommodate daily experiences.

### 6.1.2 Safety Policy

No single aspect of the work must take on greater importance than safety. Consider the field condition. Place yourself in it on a regular basis and be truthful in your evaluation of the safety quality of the area. Now imagine your children or other members of your own immediate family working daily in the same environment. Are you still comfortable with the situation? What are the things that you would do now to improve the work environment so that you would be satisfied with it? Finally, imagine your children or family members as pedestrians or sidewalk superintendents. Are you satisfied that the existing provisions will eliminate all risk to them?

It must be the policy of the Company and of each individual employed by it to always:

- Provide safe work environments
- Conduct their operations in a manner that reduces risk to themselves and to other workers
- Maintain all conditions in a way that eliminates all risk to visitors and to the public
- Eliminate risk of damage to property on and adjacent to the site

### 6.1.3 Sample Jobsite Safety Policy Notice

A notice similar to the following sample should be placed in a prominent location at the field office and in any bulletin area provided at the site for such notices. It is a statement of safety policy, and of the conduct expected to be displayed by every person on the site.

Each jobsite has its own unique arrangements and circumstances. The language of the notice can be considered a guide that can be modified as appropriate.

Like all such notices, it gets no power from the paper on which it is printed, but directly from the conduct of the Company personnel on a daily basis. Your conduct with respect to your own work and with respect to routinely policing compliance by others will be returned with equal amounts of attention.

## 6.1.3
## Sample Jobsite Safety Policy Notice

**Letterhead**

It is the policy of ( ) Company to at all times:

- Provide safe work environments,
- Conduct all operations in a manner that eliminates risk to any tradesperson,
- Maintain all conditions in a way that eliminates all risk to visitors and to the public, and
- Eliminate risk of damage to property on and adjacent to the site.

This is the fundamental responsibility of every individual on the site. All supervisors must routinely accept complete responsibility for prevention of accidents and for the safety of all work under their direction.

By contract and by law, every company on the site is obligated at a minimum to conform to the Federal Occupational Safety and Health Act, and to the laws of every entity having jurisdiction over the work.

Any company or individual refusing to correct observed safety violations will be banned from the site at least until such violations are corrected, and will be held completely responsible for all resulting effects.

There is no magic to safety. With proper attention, awareness, and cooperation by everyone, we will achieve an accident-free job, and the pride that goes with it.

( ) Company

President

## 6.2 Safety and Loss Control Responsibilities

### 6.2.1 The Project Manager

The Project Manager is ultimately responsible for the complete safety and loss control effort on the project. All reporting will be made to the Project Manager, who will:

- Administer all programs
- Disseminate all safety information
- Monitor compliance
- Evaluate potential violations
- Determine need and level of corrective actions
- Implement corrections and monitor results
- Cooperate in every way with OSHA and every organization having any jurisdiction over the work
- Ensure that all reporting is timely, accurate, and complete

### 6.2.2 The Project Engineer

The Project Engineer is responsible for providing all administrative support to the safety effort. This includes providing for and following up on all activities relating to notices, communications, investigations, and reports.

The Project Engineer may be assigned the position of Safety Officer. In that capacity, the specific responsibilities will be determined by the Project Manager as coordinated with those of the Site Superintendent.

### 6.2.3 The Site Superintendent

On small and medium-size projects the Superintendent will be the designated Safety Officer. On larger projects a Company staff individual or a jobsite individual may be assigned the responsibility as determined by the Project Manager. The specific responsibilities of the Safety Officer will be coordinated with those of the Project Engineer.

As Safety Officer, the Superintendent will administer all components of any safety program on a daily basis. Specific responsibilities include:

1. Being thoroughly familiar with all safety aspects of every operation being performed on the site
2. Ensuring that all operations are being performed properly and safely
3. Preparing reports outlining violations observed
4. Notifying the observed violator
5. Delivering all reports promptly to the Project Manager
6. Conducting "tailgate" safety meetings
7. Being familiar with and able to administer basic first aid
8. Maintaining adequate first-aid materials
9. Ensuring that the safety policy and any safety rules are posted in a prominent place

10. Enforcing compliance with all safety rules
11. Cooperating in every way with OSHA and every organization having any jurisdiction over the work

## 6.3 Safety Program

### 6.3.1 General

It is the responsibility of the Project Manager to administer the complete safety program through the Superintendent, Project Engineer, and/or designated Safety Officer. Although administered by the Company, full participation and compliance are required by every company and individual in every respect; there are no compromises.

The safety program begins during consideration of the site utilization program of Section 5.4, moves through the consideration of major site safety planning and implementation objectives, and continues through the specific considerations of the individual actions and activities going on throughout the site each day until completion and turnover of the project.

### 6.3.2 Site Safety Planning and Implementation

During determination of the site utilization program, consideration must be given to the basic logistics of safety. Items in this category include:

1. Locations and configurations of the various site services and their relative proximities:

   *a.* Field offices
   *b.* Material staging and storage areas
   *c.* Fuel storage and fuel distribution arrangements
   *d.* Traffic control
   *e.* Administrative and worker parking
   *f.* Pedestrian access
   *g.* Locations and configurations of temporary utilities
   *h.* Temporary power arrangements (throughout the construction program)
   *i.* Temporary lighting (site and building)
   *j.* Temporary heat (friendly fire)

   As the project develops, the initial determinations and arrangements for each of these items may need to be modified periodically or even routinely to keep all safety considerations adequate.

2. Specific Owner requirements that may exceed customary considerations, such as:

   *a.* Special access or security arrangements
   *b.* Special permits or other regulations
   *c.* Involvement of designated Owner safety personnel
   *d.* Special insurances or other legal considerations

3. Fire protection planning is project-specific, but will include:

   *a.* Appropriate ABC-type fire extinguishers in sizes, quantities, and locations as determined by the Project Manager for all office and storage trailers and facilities

*b.* Fire-protected storage cabinets and/or areas for vital, sensitive, or special files, materials, etc.
*c.* Establishment and maintenance of all appropriate fire protection measures in accordance with OSHA requirements for specific work operations

4. Protection of the public, including:

*a.* Signs and notices
*b.* Special walkways and traffic provisions
*c.* Barricades and other safety barriers
*d.* Fences, guardrails, and canopies
*e.* Parking, traffic, and walkway lighting
*f.* All provisions for handicap access
*g.* Security and personal safety

5. Identification of safety and emergency services:

*a.* Notify fire, police, and emergency medical facilities of the presence of the jobsite, anticipated work force, and duration of the project.
*b.* Identify all emergency room locations and know the most expedient routes.

### 6.3.3 Contractor Insurances

All Contractors and Subcontractors performing *any* work at the site must be covered by *all* insurances as enumerated in the contract documents or as specifically required by the terms of the respective subcontract agreements. This applies to individuals working as Subcontractors or Sub-Subcontractors in any capacity, as well as to all organizations participating in the project.

It is the responsibility of the Project Manager and the Site Superintendent to confirm the presence of all such insurances and secure the correct certificate of insurance from each entity on the site before allowing *any* work to proceed. There are *no* exceptions.

### 6.3.4 First Aid

The Company will provide a complete medical first-aid kit for use by Company personnel. All Contractors and Subcontractors on the site must provide and maintain their own first-aid kits in a manner adequate for the type of work being performed.

### 6.3.5 Safety and First-Aid Training

All Company personnel are encouraged to take basic first-aid training and become certified in first aid and CPR. The Superintendent, Project Manager, Project Engineer, all trade foremen, and everyone in any supervisory capacity should feel particularly responsible in this area.

Contact the local Red Cross, area hospitals, and local fire departments to identify medical first-aid training programs available, and encourage participation throughout the jobsite.

### 6.3.6 Jobsite Safety Meetings

Regular safety meetings should be conducted by the General Superintendent and/or the Safety Officer to:

- Orient all tradespeople to project requirements and objectives
- Identify specific problems on the site and the best corrective actions
- Swap or trade topics of both general and specific nature that will further the safety education and awareness of all project participants

The safety meeting series is divided between:

1. The Project Safety Orientation Meeting
2. Regular Company safety meetings
3. Periodic "tailgate" meetings

### 6.3.7 Project Safety Orientation Meeting

The Project Safety Orientation Meeting is conducted at the start of the project. It can be its own stand-alone meeting, or it can be combined with an early Superintendent's meeting with all trade foremen. Items to be reviewed at that meeting include:

1. All safety requirements of each Contractor's contracts and subcontracts
2. Company, project, and other special rules
3. The jobsite utilization program, specifically with regard to arrangements for stored materials, proper handling, etc.
4. Fire protection requirements and procedures
5. Evacuation procedures
6. Posting of all emergency phone numbers
7. Designation of each company's Safety Officer
8. Identification of all jobsite personnel with any medical training
9. Review of required first-aid and medical supplies
10. Specific jobsite precautions with respect to protection of workers and protection of the public
11. Jobsite security issues and arrangements
12. Notification to all participants that willful and/or repeated violations may be grounds for layoff or ejection from the jobsite
13. Notification of the expectation of regular participation in periodic safety meetings, and of the need for all trade foremen to conduct their own "tailgate" safety meetings periodically

### 6.3.8 Regular Safety Meetings

Regular safety meetings can be conducted as their own stand-alone arrangements. This may, however, be more effective and more easily accomplished if the meetings are combined with the regular Superintendent's meetings with all trade foremen as the first routine item on the agenda. Advantages of this approach include the ideas that:

1. Safety is a regular part of the complete production consideration. It is at least as important to the project as any material delivery or work activity.
2. Treating safety as a routine item ensures its periodic consideration without any additional amount of effort.

3. Meeting logistics are minimized. The time and effort necessary to schedule the meeting, document attendance, and distribute records is combined with that for the regular progress meeting. Time spent is therefore very efficient.
4. Regular attendance will be improved.

Meeting agenda items will include:

1. Review of any observed actual or possible safety violations, along with determination and implementation of corrective measures
2. Any planned or other changes in the site utilization program
3. Jobsite housekeeping and cleanup status
4. Review of stored materials and equipment
5. Review of past or upcoming hot work, and handling of burning and welding equipment
6. Review of temporary power provision and maintenance
7. Review of temporary heat and temporary fuel realignments and handling
8. Review of overall safety performance throughout the project

Consider combining with this meeting a tailgate meeting as described in the next section. You already have everyone's attention and participation—use it.

### 6.3.9 Tailgate Safety Meetings

The tailgate safety meeting is a 10- to 15-minute discussion that is to be conducted by the Company General Superintendent or Safety Officer with all employees and trade foremen, and by each Subcontractor's Safety Officer with its own employees. The purposes of the meetings include:

1. Dissemination of information included in the regular safety meetings.
2. Review of jobsite safety and cleanup conditions and establishment of corrective measures.
3. Discussion of a specific safety item for the training and education of all tradespeople. These items can be relevant to some observed job condition, or they can be of a random nature.

### 6.3.10 Suggested Tailgate Safety Meeting Outlines and Topics

Suggested tailgate safety meeting outlines and topics (one topic per meeting) can include:

1. Emergency care and procedures to obtain aid
    - *a.* Controlling or restoring breathing
    - *b.* Aiding a choking victim
    - *c.* Aiding a burned victim
    - *d.* Aiding a shocked victim
    - *e.* Aiding a victim with a broken bone
    - *f.* Determination of qualifications for administering first aid
2. Fire regulations at the work site

   *a.* Smoking and nonsmoking areas
   *b.* Location and use of fire fighting equipment
   *c.* Periodic check of fire extinguisher "charges"
   *d.* Storage and use of combustible materials

3. On-site accidents

   *a.* Cave-ins
   *b.* Falls—causes and prevention
   *c.* Jewelry—rings, chains, etc.
   *d.* How to manually lift loads safely

4. Chains and slings—care and proper use
5. Compressed gas cylinders

   *a.* Dangers of compressed air
   *b.* Handling and storage of cylinders

6. Use of friendly fire

   *a.* Temporary heat and fuel
   *b.* Temporary power
   *c.* Proper use of cutting equipment

7. Personal safety rules and equipment

   *a.* Hard hat
   *b.* Safety shoes
   *c.* Gloves
   *d.* Proper clothing
   *e.* Proper use of safety belts and nets
   *f.* Goggles and eye injuries

8. Safe handling of power tools

   *a.* Rip saws
   *b.* Bench grinders
   *c.* Drills
   *d.* Power-actuated tools

9. Heavy equipment—safe use

   *a.* Cranes and rigging
   *b.* Operation near embankments
   *c.* Operation in confined areas

10. Hazardous material—industrial hygiene

    *a.* Asbestos—detection and reaction
    *b.* Treatment and disposal of hazardous or controlled materials
    *c.* Carbon monoxide
    *d.* Lead
    *e.* Confined-area activity—egress, ventilation, etc.

11. Scaffolding and staging

    *a.* Erecting
    *b.* Working on it
    *c.* Use agreements

12. Ladders—types and uses

## 6.4 Sample Safety Review Checklist

### 6.4.1 General

The Sample Safety Review Checklist that follows is prepared as a convenience to aid the Superintendent and the Safety Officer in their periodic and routine reviews of the site in order to help identify possible conditions that may result in or contribute to an accident.

It is not a guarantee that all possible hazards will be identified or accommodated, nor is it a list that specifically complies with any regulatory requirements. Check with OSHA, your insurance carrier, and all appropriate authorities having jurisdiction over the work for applicable regulations.

The items listed are those that are frequently observed and easily identified.

### 6.4.2 Sample Jobsite Safety Review Checklist

## 6.5 Accident Investigation, Reporting, and Records

### 6.5.1 Reporting Requirements

All accidents must be reported. In the case of a serious accident, the Company central office must be notified as soon as practical, immediately *after* appropriate emergency measures have been taken.

For any injury to a Company employee while working, a Worker's Compensation Report Form must be completed and filed immediately with the appropriate insurance carrier.

All reporting is to be done by the Superintendent or other designated Safety Officer. It is this individual's responsibility to be aware of and strictly comply with all reporting requirements. If this function is not performed correctly, the Company will be forced to assume extreme and disproportionate amounts of liability with respect to the injured party.

### 6.5.2 Investigation Requirements

All serious accidents must be investigated. These include serious accidents to all individuals employed by the Company and any Subcontractor, and every accident involving any amount of property damage. Immediately after all appropriate emergency measures, first aid, and damage containment measures have been taken, every effort must be made to immediately:

1. Preserve physical evidence
2. Take photographs and secure other evidence as appropriate
3. Take statements from the accident victim, any eyewitnesses, and anyone who may have any knowledge of definite or possible cause(s) of the accident

If the accident is serious enough that the insurance carrier will be investigating, assist their investigation in any way they need.

# 6.4.2 Jobsite Safety Review Checklist

1. Signs, Notices, & Notifications
   a. Safety signs in place ____
   b. Emergency phone numbers posted ____
   c. Evacuation plan appropriate/posted ____
   d. Warnings & instructions to public posted ____
   e. Restricted access areas ____
   f. Exits ____
   g. No Smoking ____
   h. Electrical dangers ____
   i. Personal protective equipment needed ____
   j. Operating instructions ____
   k. Flammable materials ____
   l. Hazardous materials ____
   m. Danger areas ____
   n. Trenches ____
   o. All personnel & occupants notified to expect loud noises ____
   p. ________________ ____
   q. ________________ ____
2. Overhead Protection
   a. Entrances ____
   b. Warnings ____
   c. Construction ____
   d. ________________ ____
   e. ________________ ____
3. Hoisting Equipment
   a. Guys ____
   b. Cables & sheaves ____
   c. Turnbuckles ____
   d. Signals ____
   e. Carcover & enclosure ____
   f. Ladder ____
   g. Car arresting device ____
   h. Base barricade ____
   i. Platforms ____
   j. Clear staging areas ____
   k. ________________ ____
   l. ________________ ____
4. Walkways & Ramps
   a. Adequate construction ____
   b. Width ____
   c. Railings ____
   d. Curbs ____
   e. Slope & rise limit ____
   f. Non-slip treads & tactile areas ____
   g. ________________ ____
   h. ________________ ____
5. Ladders
   a. Construction ____
   b. Secure placement ____
   c. Cleats ____
   d. Landings ____
   e. Hand-holds ____
   f. Cages ____
   g. ________________ ____
   h. ________________ ____
6. Excavations & Trenches
   a. Shoring ____
   b. Slope repose ____
   c. Ladders ____
   d. Stockpile of excavated material ____
   e. Removal of excavated material ____
   f. Barricades & railings ____
   g. Tunnels ____
   h. Blasting arrangements ____
   i. Approved shoring designs ____
   j. Excavations properly dewatered ____
   k. Proper ventilation; free of toxic fumes ____
   l. ________________ ____
   m. ________________ ____
7. Fire Protection
   a. Storage of flamable materials ____
   b. Container markings ____
   c. Temporary heaters ____
   d. Compressed gas cylinders ____
   e. Tar kettles ____
   f. Welding equipment ____
   g. Welding operations ____
   h. Fire extinguishers (correct quant/type) ____
   i. Fire safety equipment ____
   j. ________________ ____
   k. ________________ ____
8. Openings - Walls, Floors, Roofs
   a. Perimeter railings ____
   b. Tight covers ____
   c. Flaggings ____
   d. ________________ ____
   e. ________________ ____
9. Scaffolds
   a. Construction ____
   b. Secure placement ____
   c. Railings ____
   d. Toe boards ____
   e. Rigging ____
   f. Safety lines, belts, rope guards ____
   g. ________________ ____
   h. ________________ ____
10. Stairs & Landings
   a. Adequate construction ____
   b. Temporary treads ____
   c. Clear of debris ____
   d. Proper rise/run ____
   e. Railings ____
   f. ________________ ____
   g. ________________ ____
11. Material Handling
   a. Size/bulk ____
   b. No sharp edges ____
   c. Weight limits ____
   d. Team lifting ____
   e. ________________ ____
   f. ________________ ____

12. Housekeeping
    a. Nails, debris ____
    b. Tool storage & staging ____
    c. Containers ____
    d. Clear aisles & walkways ____
    e. Clean site ____
    f. Dumpster(s) location/condition ____
    g. Proximity of waste storage to hazardous conditions ____
    h. ________________ ____
    i. ________________ ____
13. Lighting & Temporary Wiring
    a. Lighting ____
    b. Wire height ____
    c. Proper grounding ____
    d. Wire connection ____
    e. Overcurrent protection ____
    f. Extension chords in good repair ____
    g. All extension chords & temp. power recepticles using GFI's ____
    h. Temp. power closed to weather ____
    i. ________________ ____
    j. ________________ ____
14. Grounding & Electrical Equipment
    a. Correct grounding ____
    b. Ground fault interrupters ____
    c. ________________ ____
    d. ________________ ____
15. Portable & Power Saws
    a. In good condition ____
    b. Guards ____
    c. Kickback protection ____
    d. Ventilation ____
    e. Safe fuel procedures ____
    f. ________________ ____
    g. ________________ ____
16. Hand Tools
    a. In good condition ____
    b. Insulated and/or grounded ____
    c. Projectile tools ____
    d. Power actuated tools ____
    e. Operators trained in proper use ____
    f. ________________ ____
    g. ________________ ____
17. First Aid
    a. Proper kit size & contents ____
    b. Kit supply maintained ____
    c. Trained employees ____
    d. Emergency numbers posted ____
    e. Hospital routes known ____
    f. ________________ ____
    g. ________________ ____
18. Traffic Control
    a. Parking ____
    b. Speed control ____
    c. Barricades ____
    d. Separation of haul roads ____
    e. ________________ ____
    f. ________________ ____
19. Personal Protective Equipment
    a. Hard hats ____
    b. Goggles / safety glasses ____
    c. Gloves ____
    d. Respirators ____
    e. Hearing protection ____
    f. Safety shoes ____
    g. No loose clothing ____
    h. All work areas sanitary ____
    i. ________________ ____
    j. ________________ ____
20. Heavy Equipment
    a. Guards ____
    b. Warning bells ____
    c. Fueling ____
    d. Ground slope ____
    e. Rough terrain ____
    f. Cab protection ____
    g. Operator qualifications ____
    h. ________________ ____
    i. ________________ ____
21. Security
    a. Fencing ____
    b. Lighting ____
    c. Alarm systems ____
    d. Monitoring arra'ngements ____
    e. Guard service ____
    f. Target equipment ____
    g. Secure equipment practices ____
    h. Police notification procedure ____
    i. ________________ ____
    j. ________________ ____
22. Liability
    a. Release forms executed/delivered for all trades using:
       Hoists ____
       Elevators ____
       Scaffolding ____
       Equipment ____
    b. Arrange for jobsite inspection by insurance carrier ____
    c. ________________ ____
    d. ________________ ____
23. Other
    a. ________________ ____
    b. ________________ ____
    c. ________________ ____
    d. ________________ ____
    e. ________________ ____
    f. ________________ ____
    g. ________________ ____
    h. ________________ ____
    i. ________________ ____
    j. ________________ ____
    k. ________________ ____
    l. ________________ ____
    m. ________________ ____
    n. ________________ ____
    o. ________________ ____

### 6.5.3 Investigation Procedure

The investigation procedure begins as soon as all immediate danger to people and property has been brought under control, and all Company and appropriate authorities have been notified of the incident. It is to be conducted by the Superintendent or designated Safety Officer.

The purpose of the investigation procedure is to secure and confirm as many specific facts as possible, not to place blame. Once the procedure is completed, the causes and conditions can be thoroughly analyzed and determined at a later date.

Realize at the onset that the information secured at the scene may or may not be complete or accurate for any number of reasons. These can range from sincere problems in recollection or even correct perception of the event to deliberate attempts to change, conceal, or omit information. For these reasons, it is essential to:

- Keep asking who, what, where, when, how, and why.
- Work in repeated questions of the same item to the same individual for key considerations. Phrase the question in different ways and spread the repeated questions among other questions and conversation.
- Ask the same questions to as many different individuals as possible.
- Interview each individual in a separate location, keep independent answers truly independent, and eliminate the possibility of two or more individuals consciously or unconsciously "coordinating" their versions of the incident.

The procedure then will be as follows:

1. Identify all individuals who were definitely or possibly in the vicinity of the accident immediately prior to or during the accident.
2. Catalog those individuals in the order of importance:
   *a.* Those involved in the accident
   *b.* Those who may have caused or contributed to the accident
   *c.* Eyewitnesses
   *d.* Those who were in or around the proximity of the accident
3. Photograph the complete accident area.
4. Immediately summarize your own understanding of the accident, if any. Include a diagram with as much relevant information as possible, including appropriate distances and dimensions.
5. Use the *Sample Accident Investigation Report Form* of Section 6.5.4 as the baseline for your investigation, and complete it after all individuals' statements have been secured.
6. Use the *Sample Accident Eyewitness Statement Outline* of Section 6.5.5 for each of the individuals identified in items 1 or 2.
7. Upon completion, turn all accident investigation reports, photos, and information over to the Project Manager for delivery to appropriate Company personnel.

### 6.5.4 Sample Accident Investigation Report Form *(page 6.15)*

The Sample Accident Investigation Report Form that follows is an example of the report that should be completed whenever there is any significant incident.

Contact your own insurance carrier to obtain any actual Accident Investigation Report Form required by them, and use it generally in accordance with the procedures developed in this section and with any other guidelines that they may give. Review these procedures with your insurance carrier to identify any additions or modifications that he or she would consider appropriate.

### 6.5.5 Sample Accident Eyewitness Statement Outline *(page 6.16)*

The Sample Accident Eyewitness Statement Outline is to be used as an aid to help each eyewitness keep his or her statements complete, organized, and focused. It is in no way to be used to guide or force any respective statement, or as any means to add, delete, or modify any information. To the contrary, all eyewitnesses must be repeatedly encouraged to put any descriptions strictly into their *own* words.

The statement outline includes categories for all minimum information requirements and should therefore *not* be considered to be all inclusive. Each eyewitness must be encouraged to add *all* relevant information beyond the minimums outlined.

## 6.5.4 Sample Accident Investigation Report Form

Project: ______________________________ Proj.#: ______________

Company: Name: ______________________________
Address: ______________________________
Phone: ______________________________

Injured: Name: ______________________________
Home Address: ______________________________
Home Phone: ______________________________
Trade: ______________________ Position: ______________
Length of Employment: ______________

Date of Accident: ______________ Time: ________ AM PM
Date of Report: ______________ Reported By: ______________________

Type of Accident (Check One): ( ) Vehicular ( ) Personal ( ) Other
Did the injured lose any time?: ________ If so, how much (Days/Hours): ________

Was safety equipment in use at the time of the accident (hard hat, safety glasses, gloves, respirator, etc.)? ________

*(If not, it is the EMPLOYEE's sole responsibility to process his/her claim through his/her Health & Welfare Fund).*

Description of the Accident (Attach additional pages if necessary): ______________________________

What caused the Accident?: ______________________________

What has been done or will be done to correct the situation and prevent recurrence?: ______________

Who is responsible for correction?: ______________________________
Has corrective action been taken?: ( ) YES ( ) NO
If not, why?: ______________________________

Indicate streets, street names, vehicle descriptions and north arrow (use separate sheet, if necessary):

| | Injured (#1) | Driver (#2) | Driver (#3) |
|---|---|---|---|
| Insurance Carrier: | | | |
| Driver Name: | | | |
| Address: | | | |
| | | | |
| Operator Licence #: | | | |
| Vehicle Licence #: | | | |
| Vehicle Owner Name: | | | |

## 6.5.5
## Sample Eyewitness Statement Outline

Name: ____________________
Residence: ____________________
Home Phone: ____________________ Age: ______ Soc. Sec. #: ____________________

Employed By: ____________________ Position: ____________________ # Years: ____
Employer's Address: ____________________

My relationships and acquaintances with any of the parties involved in the accident are as follows:
____________________
____________________

The accident happened on (date) ______________ at (time) ______________ AM PM
The weather was (clear, foggy, rainy, etc.) ____________________
The road was (sandy, wet, dry, potholed, etc.) ____________________
Immediately before the accident:
- I saw ____________________
- I heard ____________________
- I did/was doing ____________________

During the accident:
- I saw ____________________
- I heard ____________________
- I did/was doing ____________________

Immediately after the accident:
- I saw ____________________
- I heard ____________________
- I did/was doing ____________________

I have examined the diagram attached to this statement, and have shown my position(s), the position(s) of all parties involved, and all relevant details to the best of my knowledge.
Other persons who might have witnessed or may have knowledge relating to the accident are:
____________________
I (did) (did not) make a statement to the police, opposing counsel, insurance or private investigator, etc.):
____________________ (If yes, obtain copy)
Supplementary information:
____________________
____________________
____________________
____________________
____________________
____________________
____________________
____________________
____________________
____________________
____________________
____________________
____________________
____________________
____________________
____________________
____________________
____________________

I have read these ______ pages that make up my complete statement. I have made all remarks voluntarily, to the best of my knowledge, and believe them to be true.

Signed: ____________________ Date: ____________________

Section

# 7

# Progress Schedules and Funds Analysis

# 7.1 Managing Schedules

## 7.1.1 Concepts and Section Description

"Planning" is determining the activities to be performed, along with their respective durations, and arranging them in proper sequence or otherwise defining their relationships. Placing the plan on a calendar makes it a schedule.

This section presents an overview of the primary types of schedules and their uses, suggestions to facilitate the planning function, the rights of the Contractor with respect to end dates, and the responsibilities of all parties to the scheduling effort. It then deals with the construction and use of cash-flow projections.

Throughout the section, the focus is on the logistics—the management—of the schedules, including:

- Confirming the legitimacy of the original plan
- Securing commitment from all those who must carry it out
- Determining and implementing appropriate action in all cases where performance is or may be slipping

## 7.1.2 Construction Tool or Contract Compliance?

Most specifications will include some requirement that the Contractor provide a schedule for the respective project. Primary reasons for the Owner's and design professionals' interest include:

- Assurance that a logical plan is indeed in place
- Some idea of the anticipated cash flow of the project (to help plan for their own needs)
- Documentation of the Contractor's plan and progress as a record keeping device, and as possible substantiation of or defense against claims for delay costs and consequential damages

It is becoming more common to see specifications that include increasing elaboration in the schedule requirement, apparently to assure the Owner that appropriate planning and scheduling is actually being done by the Contractor. It is an acknowledgment by the industry that too many Contractors do not normally conduct this effort adequately—and some don't do it at all—despite the fact that the effort is actually crucial to the timely, cost-effective completion of a project in a manner that will best avoid the extreme costs and liabilities associated with late completion.

Even basic scheduling efforts will satisfy most specification requirements. The point, however, is that project planning and scheduling must be done by the Company correctly and consistently because it is *the* most important function that project personnel can perform for the Company's sake. Realize now that from the schedule all else flows. Acknowledge that without an adequate scheduling effort, purchasing, correspondence, submittals and approvals, accounting for changes, and dealing with each day's decisions become random, uncoordinated, and therefore very inefficient (and possibly ineffective) activities.

The success of the scheduling effort is not so much related to the complexity of the scheduling mechanism itself or to the system selected. Whichever scheduling method is selected, success in the effort is much more related to the fact that you do it, do it well, do it consistently, and do it for the life of the project. Keep the effort foremost, and you will be rewarded with:

- Superior knowledge of your own projects in ways that you may not have thought possible
- The ability to define, track, and determine the effects of history on the project
- A stable of Subcontractors and Suppliers who will take your requirements seriously—first because they see that *you* take them so seriously
- Owners and design professionals who respond to the needs for proper actions on their part, because they see your ability to organize and present the effects of late or inappropriate action

Finally, it is perhaps most important to realize that schedules are there as tools to aid in making decisions—not to make the decisions for you.

### 7.1.3 Company Responsibilities

The nature of schedules is such that they should have been constructed *yesterday*. On many projects, certain relationships are commonly known or otherwise become apparent.

#### Example

1. The long lead item is hollow metal frames.
2. They must be coordinated with the finish hardware, purchased, submitted, and approved before fabrication can proceed.
3. The finish hardware must be finalized and approved before the hollow metal can proceed (and possibly even before it can be purchased).
4. The very first item necessary to maintain the schedule therefore turns out to be finalizing the finish hardware.

If these relationships were not known, the immediate hardware priority would be overlooked. The project would move along with those activities that are obvious for the moment (concrete, masonry), but would stop suddenly when the hollow metal frames do not arrive (and worse, when it is realized at that point that they are still weeks away...).

A comprehensive scheduling effort requires an intensive review of all construction details and component relationships to complete it properly. In such a process involving the key project participants, the obvious and subtle relationships that must be placed in proper sequence will become so disclosed. The risk of being bitten by an overlooked detail—particularly in the project's early stages—will be greatly reduced.

It is the Project Manager's responsibility to organize and finalize the development of the complete baseline schedule for each project under his or her responsibility. Contributions to the complete effort should be made by:

- All those who prepared any portion of the original bid or component estimates
- The Project Engineer and the Site Superintendent

If possible, it is also wise to involve key subcontractors or trade contractors to improve the quality of planning information. These can be the ones with whom relationships exist and who are trusted, or those who provided subcontractor and trade contractor estimates to the Company at the time of bid or proposal preparation.

Updating the schedules is *everyone's* responsibility. The Project Engineer will be responsible for actually producing the finished schedule update and transmitting correspondence, but the information necessary for the update will be generated *daily* by the Project Manager, the Superintendent(s), the Project Engineer, and everyone else with knowledge of any project effect. *These pieces of information must be recorded on the posted schedule as they occur,* for later consolidation into the finished update.

### 7.1.4 Subcontractor and Trade Contractor Responsibilities

The Pass-Through Clause (see Section 3.3.9) ties each Subcontractor directly to the complete specification requirements for scheduling as they relate to the work of each respective Subcontractor. Included here are the general responsibilities for items such as:

- Timely compliance of all work
- Providing adequate labor
- Performing the work in a manner that will not interfere with or otherwise delay the orderly sequence of work by others

Further included is the implied responsibility for the Subcontractor to meet requirements in a manner that will not cause the total contract time to be exceeded.

In addition to these very general considerations, the subcontract itself should include adequate and specific scheduling responsibilities, including ideas such as the following:

- The fact that time is of the essence to all dates and schedules.
- The acknowledgment that all schedules change continually. The Subcontractor must accordingly be aware of all *current* schedule requirements as they may have changed, and comply with them in every respect.
- The agreement to adjust manpower, equipment, overtime, and Saturday, Sunday, and holiday work as necessary to meet all schedules.

Beyond these stipulations, the subcontract should also contain an adequate *Acceleration Clause.* This language would give the General Contractor the right to accelerate the work of any particular Subcontractor by specifically directing the addition of labor, equipment, overtime, or Saturday, Sunday, and holiday work whenever it becomes apparent to the General Contractor that a respective portion of the work is not likely to be completed on time or as promised. The clause can then go on to state that:

- If the acceleration is not due to the fault of the particular Subcontractor, he or she may be reimbursed for the acceleration costs.
- If it is due to the fault of the Subcontractor, the Subcontractor will remain responsible for all acceleration costs.

- However, in either case, the Subcontractor *cannot refuse* to accelerate as specifically directed.

Finally, some condition (such as tying nonperformance directly to the liquidated damages provision of the general contract) should be considered to be assigned to the respective Subcontractor's failure to accelerate when so directed.

With all these rights, responsibilities, and tools, the management of the Subcontractors' portions of scheduling responsibilities will boil down to the amount of direct and specific effort made by the Project Manager, Project Engineer, and Site Superintendent(s) in involving the Subcontractors *daily*.

## 7.2 Schedule Types and Uses

### 7.2.1 General

Schedule types vary greatly in their:

- Simplicity or complexity
- Levels of detail
- Ability to actually display the plan (How did we decide to place that activity there...?)
- Visibility (Can we see and understand the information?)
- Record-keeping ability (What actually happened?)
- Ability to display cause and effect

Each schedule type has its purpose. Each is strong in one or a few of the ideals listed and weak in others.

Schedule selection criteria should be based upon the actual needs of the respective parties. Examples may include the following:

- Building committees or the Owner's Finance Committee may only want to see a bar chart to get a general sense of the project.
- Although you've just selected the most expensive scheduling software on the market, are you really interested in "resource leveling," particularly if you subcontract most of your work? If so, do you really think that your project people will be conducting the operation consistently for the life of the project, and you will be able to manage appropriate Subcontractor responses effectively?
- Do you really need to break your schedule into 2000, 1000, or even 500 activities? Are you trying to micromanage the work from your office miles away from the site? What do you think the probability is of conducting a greatly detailed sequence of activities exactly as planned—one year from now?

These general ideas must be decided upon before the scheduling method—or combinations of scheduling methods—are selected. The description that follows will help guide those decisions.

The information in this section is a guide. It is not sufficient by itself to develop complete proficiency in any particular scheduling method. For that, more time and effort will be required, as well as the complete materials needed for this purpose.

## 7.2.2 Bar Charts

### Description

1. The simplest of all methods, activity descriptions are placed in a column on the left, along with other desired relevant information (budget cost, for example).
2. A calendar is placed along the top of the chart, extending for the duration of the project.
3. A line or "bar" is placed alongside each activity's description in the time area to correspond with the calendar above.

### Example

**7.2.2**
**Example Bar Chart**

```
DAY                1  2  3  4  5  6  7  8  9 10 11 12 13 14 15 16 17 18 19 20 21 22 23 24 25 26 27 28
------------------------------------------------------------------------------------------------------
Foundation         ---------------------------
Masonry                                       -----------------
Roof Framing                                                    -----------------
Roof                                                                             -------
Slab On Grade                                             ------
MEP Underground                                  ----------
Complete MEP                                                                           ----------------
Subst. Completion                                                                                      ◆
```

### Advantages

1. It is simple to prepare.
2. It displays anticipated time frames of major activities visually.
3. It is easy to understand; can be a good "communicator" to large groups such as building committees.
4. It is historically accepted—used by most Superintendents and project personnel since the beginning of time; accepted by many design professionals and Owners.
5. Activities are normally listed in order of specification section. It is easy to correlate with the Schedule of Values and other payment records.

**Disadvantages**

1. It is most often oversimplified in its approach, not providing enough detail to use as an actual management tool.
2. It does not display the activity relationships. The schedule preparer *intuitively* considered these, but the logic is lost. This problem is emphasized because the activities are normally listed in order of specification sequence, as opposed to any logical sequence.
3. It is seldom updated or corrected.
4. "Updating" can only record a greatly simplified version of history—only when a respective activity actually started and ended. It cannot:

   - Display cause and effect
   - Forecast effects of current (good or bad) activity status
   - Accommodate changes in any way

5. Any efforts by project people to reschedule or accommodate any changes at all from the original program are completely left to the memories of those conducting the change.
6. You have no ability at all to communicate changes and their effects—not just to the Owner or Architect, but to your own people.

**Conclusion.** A bar chart can be a quick, convenient *preliminary* planning tool that can give a good idea of the general project parameters at the onset of a project, and as a first guide to a more useful scheduling method. It is also a good presentation vehicle that can place complex relationships into understandable form for large groups. It is not, however, an acceptable tool to manage a project by itself.

### 7.2.3 CPM/PDM

**Description.** CPM, or critical path method, has become an overused buzzword in the industry. It is usually (but not necessarily) a computer-generated schedule that accommodates each activity's start and completion points, duration, dependencies on previous activities, and relationships with succeeding activities. The activity relationships are generally defined into the computer, and the computer is given the job of figuring out the resulting "network"—where each activity will actually fall in its complete set of relationships and on the calendar. As the job progresses, information regarding actual start and completion dates, durations, logic corrections, changes, additions, and deletions are added to each activity's information, and an update or "reschedule" is generated by the computer. Displays are either "activity-on-arrow" or "activity-on-node."

PDM, or precedence diagramming method, is an "enhanced" version of CPM that adds an attempt to provide probabilities of the success of various baseline and update schedule scenarios. PDM may still be used in other industries such as aerospace, chemistry, and so on, but has generally been abandoned for construction because of its complexity that is wholly unnecessary for our needs. For this reason, the remaining discussion will focus on CPM only.

**Example**

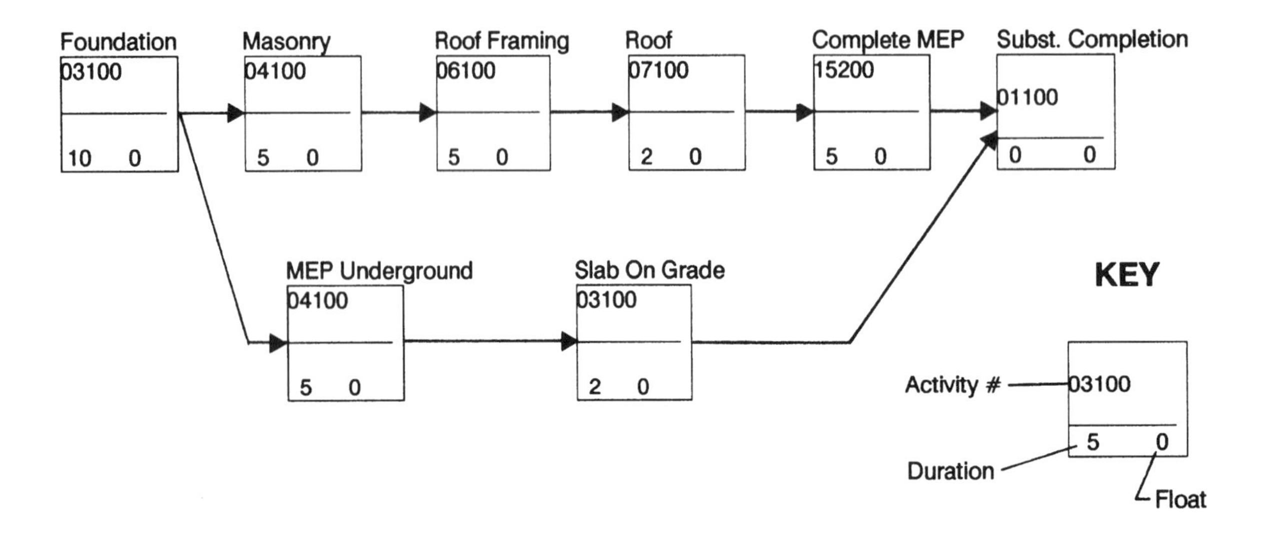


**Advantages**

1. It can be developed to any level of detail. Originally developed as a method to accommodate extremely large or complex schedules (building a missile system, for example), CPM can handle more than adequate levels of detail needed for any type of construction project. The schedule size is limited only by computer memory, which is becoming cheaper every day.
2. It can accommodate any number of changes to the program—such as simple corrections to logic, or change orders, delays, etc.
3. It will display all activity relationships that define the original and changed schedules. The logic of all "before and afters" is left intact.
4. There is great flexibility for adjustment throughout the baseline schedule development.
5. Computer programs can provide many types of reporting to managers, which can be tailored to exactly how each manager wishes to run his or her projects.

**Disadvantages**

1. It requires a computer, specialized software, and basic computer familiarity. Specialized training used to be a significant drawback. These days, however, the software is very straightforward in its use, and very "friendly." A little initiative and a dose of faith (if you are not familiar with computers) should be all that is necessary to be trained or to train yourself in a reasonably short period of time. This "disadvantage" is becoming less so each day.
2. It has very poor visibility. Unless combined with the "logic diagramming method" as described below or something similar, CPM diagrams themselves display no relationship to time. An activity requiring 5 days to complete will take up as much space on the diagram as one requiring 100 days to complete.

   Even the most accomplished CPM purist can therefore not get any "sense" of where the schedule really is by simply viewing it. The information must be studied, clearly understood, and visualized by the reviewer. All analysis is accordingly left to straight consideration of tedious lists of dates. This is by far the most serious disadvantage for any manager or communicator.
3. It must be updated completely—from start to finish and in all its detail—each time that it is updated. This can be a large task, which may become compromised in practice.
4. Each update generates essentially an entirely new schedule. The set of original activity targets, dates, and milestones can be too easily (if not conceptually) lost after just one or two updates if the manager is not careful to take definite steps to prevent this from happening.

**Conclusion.** CPM logic is the basis for all effective scheduling methods. Its earlier forms used on computers with limited graphics capacity were more prone to the "visibility" disadvantages listed. Later forms combine the CPM logic with the improved visibility of logic diagrams. These combined forms, described in the next section, are the most effective as management tools.

## 7.2.4 Logic Diagrams

**Description.** The logic diagram became the natural extension of CPM as computer graphic capability became easier and economical, in order to provide the visibility that was badly lacking.

It is simply the capability of taking the schematic, unscaled activity "nodes" of CPM, scaling them with respect to time, and placing them in their proper places under a calendar. The result appears visually to be a time-scaled bar chart with the dependencies drawn in. The bars themselves are generally "optimized" in their display—fit onto the paper wherever there is space available. The result is that the activity placement is by schedule logic (as opposed to specification section, as in the bar chart).

The logic diagram is therefore only a drafting convention. The schedule itself remains a true CPM schedule.

**Example**

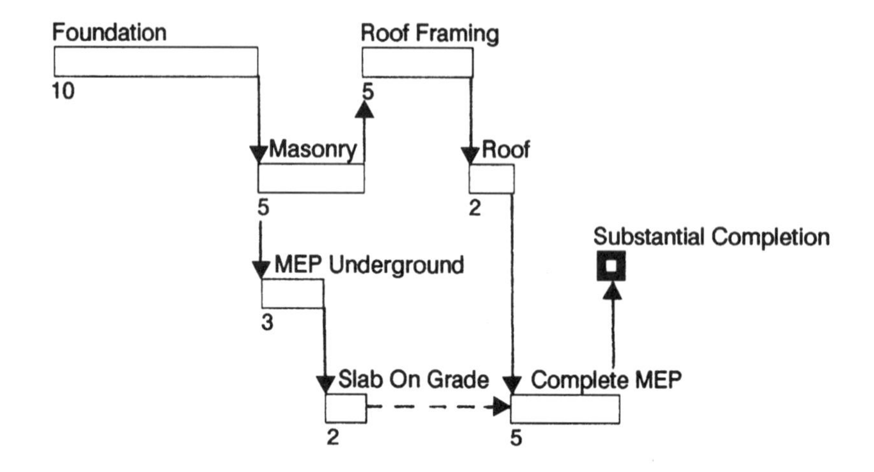


**Advantages**

1. All the advantages of CPM apply, as listed in Section 7.2.3.
2. There is greatly improved visibility—and understandability—of all schedule information, much better than even the bar chart.

**Disadvantages.** The problems listed for CPM under the following items remain:

- 1—computer requirement
- 3—need for complete update
- 4—potential practical loss of previous information

**Conclusion.** Logic diagrams are so significant in their contribution to the development, implementation, and management of CPM schedules and now so available in their software that no CPM schedule should ever be used without them. With the possible exception of the MOST method described in Section 7.2.5, the CPM/logic diagram combination is one of the most effective management methods available and must be seriously considered. Its selection as opposed to the use of MOST will be determined by the manager's preference, needs, and abilities.

## 7.2.5 MOST

**Description.** "MOST," or management operation system technique, is a scheduling method originally developed as a companion to PERT and CPM before computer graphics become available, specifically to add visibility to CPM schedules. As developed for construction, it quickly evolved into a stand-alone system with major advantages.

Before the availability of computer graphics that could generate logic diagrams, MOST began as a manual drafting of what the industry later dubbed the "logic diagram." The MOST baseline schedule therefore appears nearly identical to the logic diagram.

MOST differs from all other scheduling methods in its unique ability to record all project effects in a way that is almost unbelievably clear, to pinpoint complete accountability for all project effects, and to reschedule without redrafting or replotting. This last feature is the mechanism that allows the MOST schedule system to be used effectively without any computer. Updating it is very easy, and is accomplished in less time than for the other methods.

#### Example

### 7.2.5
### MOST Example

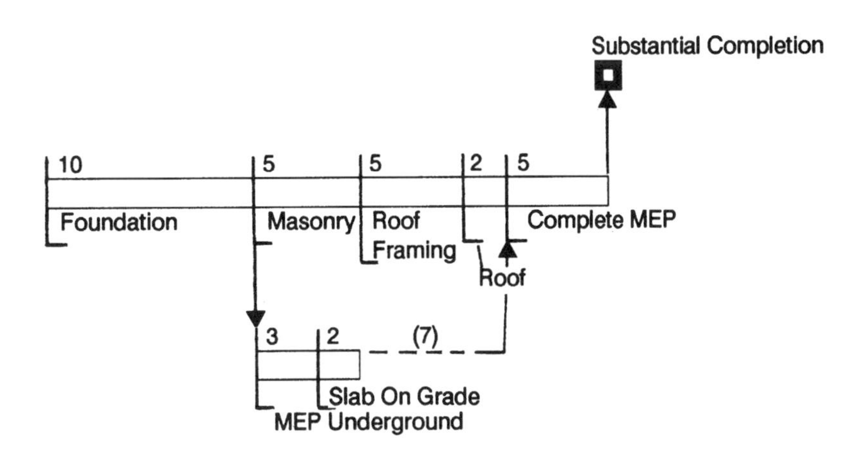


#### Advantages

1. It has all the advantages of the logic diagram as described.
2. It needs no computer. A computer is effective during the development of the baseline schedule because of the convenience for modification. However, a computer is not at all necessary.
3. It is possible to reschedule without redrawing or replotting. Reschedules and updates do not require that the schedule be redrafted. This feature allows all updates to be conducted manually.
4. The entire schedule or only the portion of immediate interest need be updated.
5. It always retains all original activity plans and milestones visibly. The current status is always placed against the *original* program to display exactly what is going on.

6. It greatly simplifies the otherwise complex history of detailed effects on activities and sequences. It keeps names pinned to effects, and it specifically quantifies the effects.

**Disadvantage.** The updating method will not computer-generate reports.

**Conclusion.** Probably the most effective combination of schedules is to:

1. Use a computer with CPM/logic diagramming abilities to develop the original baseline schedule and, if a plotter is available, to plot the baseline schedule to save drafting time.
2. Once the final baseline is determined and either drafted or plotted, use the MOST technique for all updating, reporting, and presenting.

## 7.3 Schedule Preparation and Development

### 7.3.1 General Procedure

1. The Project Manager will direct the development of the baseline schedule and oversee its periodic updates. The Project Manager, the Project Engineer, or the Company Scheduler can construct the actual document, and it will be done so with information coordinated from many sources, including:
   *a.* The project team's sense of specific logic sequences and the direct application of relevant experiences
   *b.* The bid or proposal recapitulation, including detailed cost estimates
   *c.* Schedule considerations in the original estimates
   *d.* The plans, specifications, bid documents, and referenced standards
2. As the schedule draft is developed, all activities must be studied for their specific relationships with contiguous work. Each activity must be identified or coordinated with:
   *a.* Its specification section number and description
   *b.* The specific details for the specific project (subtle differences can dramatically alter "familiar" sequences)
   *c.* Complete, accurate item descriptions
   *d.* All submittal or approval constraints
   *e.* The extent of required submittals, anticipated time to prepare and submit
   *f.* The estimated time for design coordination and approvals
   *g.* Material fabrication and delivery times after receipt of approved submittals
   *h.* Erection and installation times
   *i.* Specific erection and installation sequences and constraints
   *j.* The need for any "built-in" items by others before or with the item itself
   *k.* Items required to be installed before or coordinated with
   *l.* Manner of erection and installation—steady sequence or intermittent
3. To the complete extent possible, involve key Subcontractors or trade Contractors as early as feasible in the rough draft development. At the very least, use their specialized experiences if the bid packages have not yet been sublet, or incorporate actual subcontract commitments if they have been. This will help improve the accuracy of the decisions made with respect to the consid-

erations of item 2. Wherever possible, allow for slightly more time than a particular Subcontractor says he or she "needs" to complete an activity. This can be tightened up later in the final schedule development if need be.

4. Identify and consider the timing of all key items, including:

   *a.* Contract start date
   *b.* Actual project start date (if different)
   *c.* Anticipated shut-down periods
   *d.* Specific contract, site, major sequence, and weather constraints
   *e.* Characteristics of major components, such as:

      (1) Delivery and erection of superstructure
      (2) Delivery of switch gear, light fixtures, etc.
      (3) Delivery of key HVAC equipment
      (4) Delivery of unusual components that dramatically affect sequences (for example, spandrell glass roof)

   *f.* Project milestones, such as:

      (1) Roof complete
      (2) Building weathertight
      (3) Permanent power
      (4) Permanent heat
      (5) Substantial completion
      (6) Punchlist period
      (7) Turnkey date

5. As the draft nears completion, conduct a preliminary confirmation meeting with all key players signed on to the project by then:

   *a.* Review the details of the entire draft with everyone together.
   *b.* Negotiate all sequences to the agreement of everyone, and secure confirmation of their respective intentions to proceed as coordinated.
   *c.* Take attendance. The resulting final schedule, along with an appropriate confirming letter of transmittal, will normally be sufficient to serve as the record of the meeting.

6. Finalize the complete schedule and distribute it in accordance with Section 7.4.

### 7.3.2 Schedule versus Contract End Dates

The schedule and contract end dates will not necessarily be the same. The contract will require completion of the schedule within a certain number of working or calendar days, but this does not require that your schedule be *extended* to this complete period if it does not need to be. The two situations then will be:

1. A schedule draft exceeding the contract time
2. A schedule requiring less than the contract time allowed

**A schedule draft exceeding the contract time.** In practice, if the first baseline schedule draft exceeds the contract time period, it must be massaged and shortened to fall within the allotted contract time. If efforts to accomplish this are foiled by some specification requirement that is beyond your control, an extension of the contract time (along with appropriate compensation) can be in order.

Examples of this kind of effect include:

- Specified equipment that is not available to be delivered to the project within the required time period
- Areas of the project that will not become available to the Contractor to work in sufficient time during the contract period
- Unusual sequences or events that could not have been reasonably anticipated at the time of bid

The Owner and the design professionals had an obligation, throughout the preparation of the contract documents, to provide a specification that is consistent—one that is *achievable* within the allotted time, notwithstanding the fact that there may be several ways to build the same project. In addition, it is the design professional who carries the *implied warranty* that this is the case. If subsequently it is discovered that such performance is impossible, it will be the designer who will bear the responsibility. Refer to Section 3.3.14, *Impossibility and Impracticability,* for related discussion.

**A schedule requiring less than the contract time allowed.** If it is reasonably demonstrated that the schedule is achievable in less time than that allowed by contract, the Contractor has a right to so complete it.

If there is *very* clear and complete (and fairly elaborate) language in the contract as to the Owner's lack of ability and/or intention to take the project any sooner, there may be a basis for the Contractor to "carry" the project unoccupied to a certain degree. But this case is extremely rare, and even then suspect.

If the true schedule that is being used to manage the project is shorter than the contract time, the Owner, design professionals, Subcontractors, or anyone else may not interfere with it. They will be just as responsible for a delay in the *schedule* end date as for a delay in the contract date.

Know this condition, and tolerate no argument to the contrary. Owners who have the delay-claim experience (many public entities, for example) will argue energetically that your schedule *must* correspond to the contract time. Realize that these points are attempts to cleanly sidestep complete responsibility for at least the time difference between the schedule end date and the contract date if a delay situation develops.

## 7.4 Baseline Schedule Distribution and Final Confirmation

### 7.4.1 Final Baseline Schedule Confirmation

By the time the schedule is ready to be released (assuming the effort has been conducted as early as possible), many (but not all) of the major Subcontractors and Suppliers should have participated. The Owner and the design professionals will not have participated (unless serious problems became apparent that required design changes), and many major and minor Subcontractors and Suppliers for various bid packages will not yet have been signed on. For these items, extreme care should have been taken to verify the validity and reasonableness of all planning information, so that problems and surprises with new constraints disclosed when those items are finally purchased are minimized.

The baseline schedule should accordingly be confirmed to the fullest extent possible before eventual submission to the Owner as the complete document against which all project performance is to be measured. Accordingly the sequence will be as follows:

1. Transmit the schedule to all who participated in its development in order to secure final confirmation of all information. Tolerate only the most serious pieces of new information that cannot be modified at this point if anyone wishes to change previously coordinated information.
2. Transmit the final confirmed schedule to the Owner, first in accordance with any specification requirement, and then with any Company requirements.
3. Use the finalized schedule as the latest specific subcontract or trade contract requirement for each new Vendor signed on to the project.

### 7.4.2 Sample Letter to Subcontractors and Suppliers Regarding Baseline Schedule Confirmation *(page 7.16)*

The Sample Letter to Subcontractors and Suppliers Regarding Baseline Schedule Confirmation is to be used to transmit the baseline schedule to all those who participated in the schedule's development or have otherwise been signed on to the project as of the schedule release date. Its purpose is to secure positive confirmation from all key project participants that the specific performance requirements have been directly considered and that they agree to adhere to them in every respect.

### 7.4.3 Sample Letter to Subcontractors and Suppliers Regarding Baseline Schedule Final Release *(page 7.17)*

The Sample Letter to Subcontractors and Suppliers Regarding Baseline Schedule Final Release is to be used to transmit the final schedule to all parties after all confirmation efforts have been completed. It is therefore to be used to send the final schedule requirements to:

1. All Vendors who have appropriately responded to previous coordination efforts
2. All Vendors who have not responded to the Sample Letter to Subcontractors and Suppliers Regarding Baseline Schedule Confirmation of Section 7.4.2 by the date required
3. All Vendors signed on to the project after the schedule was confirmed

## 7.5 The Cash-Flow Projection

### 7.5.1 General

The cash-flow projection is the monthly and cumulative description of the value of the work that is anticipated to be completed within each respective payment time period. It relates the approved Schedule of Values directly to the finalized baseline

## 7.4.2
## Sample Letter to Subcontractors and Suppliers Regarding Baseline Schedule Confirmation

**Letterhead**

To: 1) ____________________
2) ____________________
3) ____________________
4) ____________________
5) ____________________
6) ____________________
7) ____________________
8) ____________________
9) ____________________

Date: ____________________

Project: ____________________

____________________

Project No. ____________________

Subj.: Confirmation of Baseline Construction Schedule

Gentlemen:

Attached are two copies of the Baseline Construction Schedule finalized as a result of the schedule coordination meeting conducted on ________________, 19____, indicating your general items of work and their relationship with the project. Note that the omission of any items required by your respective contracts does not relieve you of the requirement.

Please review the schedule, specifically considering all that is necessary to achieve the performance results indicated for the complete project, including:

* Submittal dates and durations.
* Material fabrication and delivery times.
* Activity durations.
* Dependencies (work required before your work can proceed).
* Logic (correct sequences).
* All other relevant considerations.

If modification to the schedule as it is represented is necessary, note same in red and return one copy to my attention for approval. If the schedule is acceptable as it is, please so confirm by signing one copy noting "approved," and return it to my attention.

Your response is required by ____________________, 19 ____.

Thank you for your cooperation.

Very truly yours,

COMPANY

______________________________
Project Manager

cc: File: Baseline Sched. w/att.
CF

## 7.4.3
## Sample Letter to Subcontractors and Suppliers Regarding Baseline Schedule Final Release

**Letterhead**

To: 1) ______________________
2) ______________________
3) ______________________
4) ______________________
5) ______________________
6) ______________________
7) ______________________
8) ______________________
9) ______________________

Date: ______________________

Project: ______________________
______________________

Project No. ______________________

Subj.: Baseline Schedule
Final Release

Gentlemen:

Attached is the final Baseline Construction Schedule for the project.

This schedule will be periodically updated by this office to reflect current progress, and to identify changes to the program that may become necessary to maintain critical dates.

Per your subcontract, it is your responsibility to be aware of all current schedule conditions *as may be so changed from time to time*, and to comply with them in every respect. Failure to review all current information and to be aware of all current requirements will in no way relieve you of the responsibility for the information, obligations, and performance requirements.

Please acknowledge receipt of this schedule by signing below, and returning this letter to my attention.

Your response is required by ______________________, 19 ____.

Thank you for your cooperation.

Very truly yours,

COMPANY

______________________
Project Manager

cc: File: Baseline Sched. w/att.
CF

construction schedule to result in a schedule of planned billings for the life of the project.

Its accuracy is directly dependent upon:

- The accuracy of the general Schedule of Values
- The degree of direct correlation of the general Schedule of Values with those of each major Subcontractor and Supplier
- The accuracy of the baseline construction schedule
- The degree of direct correlation of the construction schedule with the general Schedule of Values (and the corresponding lack of subjectivity in relating the two)

If the accuracy and direct correlation of all these documents can be maintained:

- The cash-flow projection will be realistic to begin with.
- The relation of cash progress with schedule progress will be accurate.
- Subsequent cause-and-effect and even consequential damage analysis will remain valid.

## 7.5.2 Preparation

Preparation is most straightforward if care has been taken to relate directly the baseline schedule to the general Schedule of Values. This itself is more readily achievable if each has been related directly to the project specification in the first place.

In any case, the complete procedure boils down to:

1. Assigning an appropriate cost to each scheduled activity
2. Distributing that cost appropriately within the activity duration (as recognized to be billable)
3. Relating those distributed costs to the respective payment periods in which each falls

The procedure is essentially the same, whether a computer schedule is used or the information is calculated manually. Computer programs will often have a mechanism to allow assignment of cost to an activity, and also allow some degree of control as to the placement of cost in an either distributed or point-cost manner. If not, the computer schedule plot can be used as it is, with the cost-plotting procedure conducted manually.

In either case, the critical issue is keeping complete and accurate notes as to the actual estimates, determinations, and procedures used throughout the preparation exercise in order to substantiate your work if the projection should ever be questioned at a later date.

## 7.5.3 Procedure

The procedure described is to construct the projection manually. The principles are identical to those applied to any computer program, and should be treated accordingly.

1. Assuming a time-scaled schedule, draw vertical lines through the schedule at each payment period (presumably months). Note that this step is not necessary with computer-generated cash projections.
2. Assign costs for all schedule activity items and material deliveries.
   *a.* For each activity and material delivery item, insert the total cost estimate as your reference for eventual distribution along the respective activity.
   *b.* Keep this total cost distribution as closely related as possible to the approved Schedule of Values.
3. Distribute activity costs as appropriate for each specific item.
   *a.* For on-site activities it will generally be most appropriate to distribute the total activity's cost evenly over the duration of the activity.
      (1) With the exception of unusual or very long (and possibly oversimplified) items, this method will be sufficient for the purposes of the projection.
      (2) If there is a significant reason why a relatively even distribution of costs will not be appropriate, consider breaking the activity into those components that define the detail more correctly and assign the separate costs accordingly.
   *b.* For computer programs it will normally be sufficient to simply indicate this distribution instruction. For manual preparation, simply prorate the percentage of cost to the same percentage of the actual activity time falling in each payment period.
4. Distribute material delivery costs.
   *a.* Most contracts provide that materials will be paid for only when delivered to the site. If the prescribed conditions are different, so be it.
   *b.* While it is true that other payment conditions may apply in problem circumstances, they are normally there to deal with special situations that may develop. They accordingly have no place in a projection that does not directly anticipate such a problem. If you are aware of payment conditions that allow for off-site payment, your schedule should have a "delivery/storage" bar or item that specifically shows it. It can then be treated in the same manner as all other delivery items.
   *c.* Because of the foregoing conditions it will then be appropriate to assign 100 percent of the complete material cost for the item to the last day of the material delivery schedule time bar or item. In other words, all of it will be billable when all of it arrives at the jobsite.
5. Consolidate all costs. Total all costs for each payment period, and enter the value at the bottom of the column.
6. Apply overhead, supervision, profit, and all other direct and soft costs either distributed evenly over the entire schedule or as otherwise appropriate. This step can be omitted if these cost factors had already been considered in the preparation of the individual activity costs as "complete" costs in the preceding steps.
   *a.* Some costs such as "jobsite overhead" are more likely to be accurate if spread evenly per month (trailer rental, for example).
   *b.* Some costs are more appropriate to be period-specific (bond premium costs falling in the first month, for example).
   *c.* Some cost categories can go either way. (Profit can be treated as applying strictly to the same percentage as work complete, or it can be more directly related to the specific bid packages.)

7. Add step 6 overhead, supervision, and profit to the total direct costs of step 5 to arrive at the total receivable values for each payment period. The cash-flow projection itself is not complete.

The simplified illustration given in Section 7.5.4 shows the specific steps as described. This illustration is complete; more detailed schedules only have more items.

## 7.5.4 Example Cash-Flow Preparation Worksheet

**7.5.4**
**Example Cash-Flow Projection Worksheet**

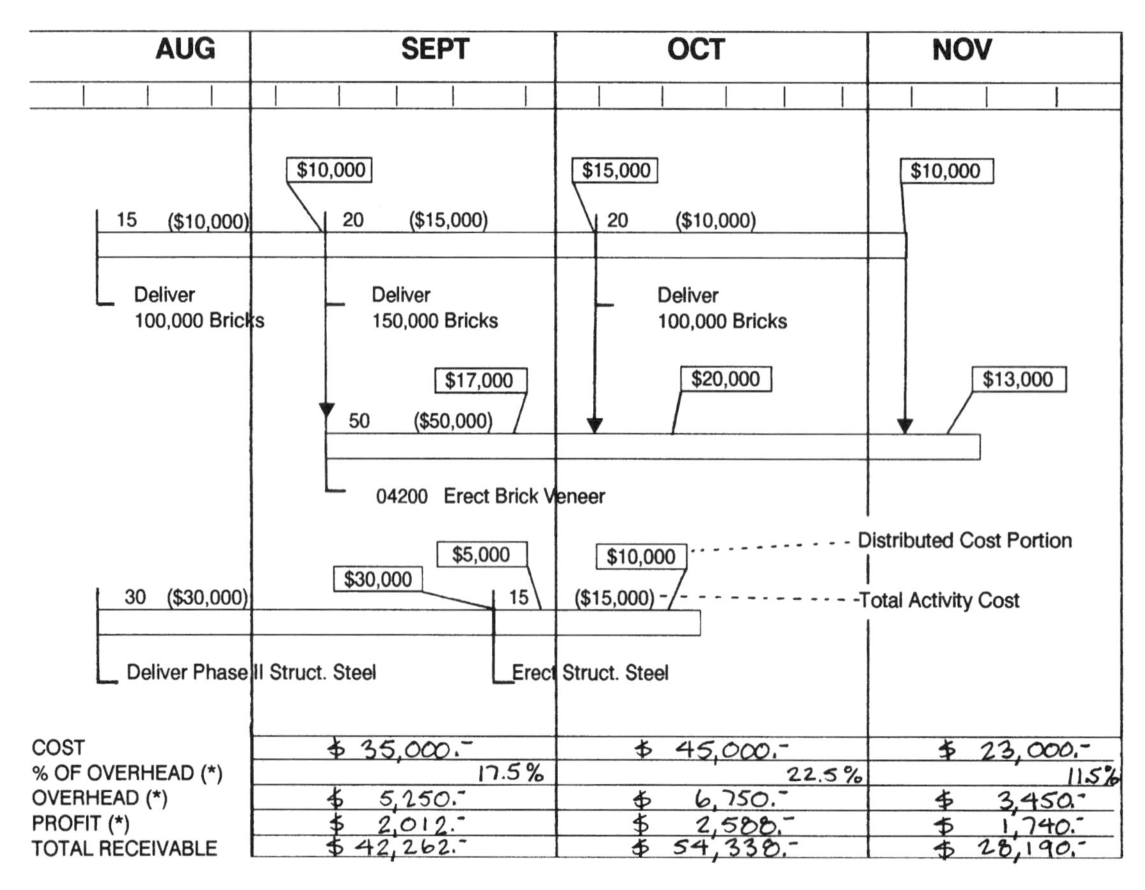


| | SEPT | OCT | NOV |
|---|---|---|---|
| COST | $ 35,000.- | $ 45,000.- | $ 23,000.- |
| % OF OVERHEAD (*) | 17.5% | 22.5% | 11.5% |
| OVERHEAD (*) | $ 5,250.- | $ 6,750.- | $ 3,450.- |
| PROFIT (*) | $ 2,012.- | $ 2,588.- | $ 1,740.- |
| TOTAL RECEIVABLE | $ 42,262.- | $ 54,338.- | $ 28,190.- |

(*) For this example:
Total Project Cost is $200,000
Overhead is $30,000
Profit is 5%, and distributed evenly

### 7.5.5 The S-Curve

This section describes the creation of the S-curve and includes basic direction on its uses and purposes. You should be aware that very elaborate analyses of total project performance and the ability to demonstrate many direct cause-and-effect relationships are possible with the S-curve, but these detailed developments are beyond the scope of this Manual.

The S-curve is a cumulative plot of the cash-flow projection as developed in Section 7.5.3. In order to create it, simply:

1. Cumulate each month's total value with the combined values of all preceding periods.
2. Plot on graph paper to create the S-curve.

In the example given in Section 7.5.4, and assuming September in the example is the project's first period, the information would be as follows:

| | Period 1 September | Period 2 October | Period 3 November |
|---|---|---|---|
| Total receivable | $42,262 | $54,338 | $28,190 |
| Cumulative totals | $42,262 | $96,600 | $124,790 |

**Example S-Curve**

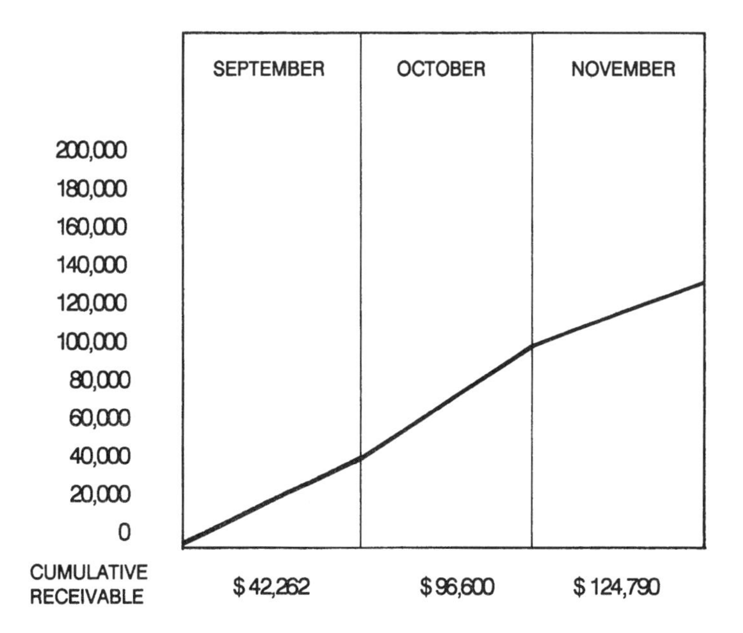

The S-curve plot for these three periods would look like the following:

In this simplified example the "curve" is abrupt, with no distinct shape. In an actual, detailed analysis of an entire project spanning many time periods, the curve would begin shallow, steepen during the middle periods, and then level off again in the ending periods. This is a common characteristic that basically represents slower "start-up" activities in the beginning, the project maturing into its production phase of high activity, and closing off with slowing activities and with finishes that are not as high in cost relative to the complete project. It is this characteristic that will cause the actual curve to approximate the shape of an S. An actual S-curve for an entire project may look generally like the following:

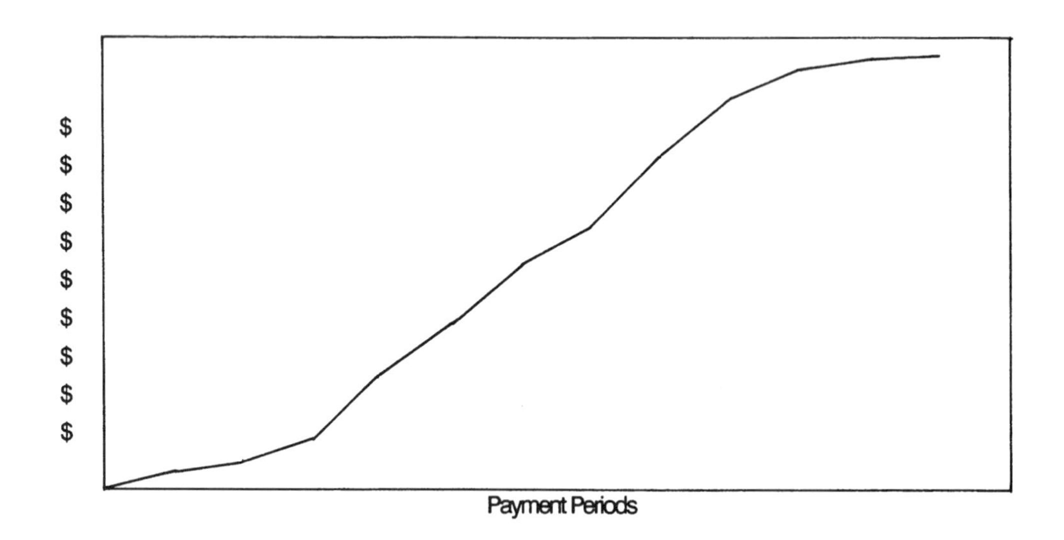


As a footnote, because this shape is so characteristic of most conventional schedules, if you should note that the shape of one of your curves is dramatically different, it may be an indication that the schedule itself is not realistic or is inaccurate. There may turn out to be a perfectly good explanation, but in such a case the S-curve analysis will at least have forced a review to confirm the validity of the original schedule. Most often, however, such a reconsideration would disclose an error.

### 7.5.6 Comparison of Actual to Projected Cash Flow

The first use of the S-curve is to allow planning for the cash flow for each project as specifically related to planned progress, and then monitoring such performance on an ongoing basis. As part of that function, actual applications for payment should be compared to those planned—as they are occurring. If this simple procedure is conducted for all projects on an ongoing basis, total Company cash flow becomes an item that is planned and managed, not something that becomes known as it happens (for better or worse).

In order to do this, plot the values of each requisition (contract value without approved change orders) on the same plot as the original S-curve. The relationships will immediately become apparent:

- If you are reporting the project "on schedule," the actual requisitions should be very closely paralleling the projected curve.
- If the project is being reported slightly behind schedule, the curve should be falling off slightly.
- If the project is being reported ahead of schedule, the actual payment line should be falling above the planned curve to the same degree.

If this direct one-to-one relationship between the reported schedule activity progress and actual billings is not apparent, something is wrong with one or more of the following:

1. The baseline schedule is inaccurate.
2. The cash-flow projection had been prepared incorrectly.
3. Schedule progress reporting is not correct.
4. Billings are not being conducted correctly.
5. Billing *approvals* by the Owner or the Architect are not correct.

If the schedule is truly being used to manage the project, the situation as it exists will become apparent to management as soon as it is occurring—in plenty of time to do something about it before it gets out of control.

### 7.5.7 Comparison of Cash Progress with Activity Progress

The S-curve can be very useful in validating the original schedule and supporting the schedule updates. The basics of the subject are reviewed here while the very detailed development of the techniques is beyond the scope of this Manual.

**Curve shape and slope.** The first analysis has to do with the shape of the curve itself. The slope, or steepness, of the curve is actually the rate of progress of the work. The steeper the slope, the higher the billings, the more work is being represented to be complete.

The S shape is the result of the characteristic work flow of a project which includes:

1. A mobilization and start-up period (shallow curve slope)
2. Work increasing and developing to the "pace" of the project (steepening curve slope)
3. A general slowing down of the rate of progress, and work being done on finishes—work that is lower in cost relative to the complete project (flattening curve slope)

As mentioned previously, if the curve does not assume the characteristic S shape, there is a good chance that something may be wrong with the planned schedule. Such common problems include:

1. The curve begins too steeply, indicating that too much work is anticipated too soon in the very beginning of the project.
2. The shape varies greatly among payment periods in the job's pace cycle, indicating billings that are jumping around when progress should actually be steady.

3. The curve ends too abruptly instead of flattening out. This can mean that you are expecting the project to run to its end at the rate of its pace cycle and then just stop—possible but not probable.

There may be good explanations for these effects (such as coincidental delivery of a large number of expensive equipment items that causes the projected value of a certain period to jump dramatically). The point, however, is simply that if the curve varies dramatically from the expected shape, you must either confirm that there are appropriate reasons, or fix the mistake immediately.

**Comparison of planned curve with actual curve as a means of validating the original schedule itself.** In a delay situation, the original schedule will become the subject of an extreme amount of criticism. It will be accused of being too aggressive and not properly prepared in the first place. In the worst conditions, it might even be accused of being intentionally prepared in an overly aggressive manner, specifically to maximize future delay claims.

If such a situation develops in which the project was delayed, the cause of the delay is eventually resolved, and work resumes, the curve can go a long way in substantiating the validity of the planned program.

In such a delay situation, the planned curve for the period will typically be steep (indicating a good amount of planned progress), and the effects of the delay will flatten the curve of the actual payments (indicating that less work was done than planned). When work resumes, the actual curve should then approximate the original curve of the preceding delay period. If it does, the confirmed actual rate of progress (steepness of the actual curve) that matches planned curves for the same kind of work will directly confirm that the original planned work was absolutely achievable in the first place. It will nail down the fact that the planned schedule was realistic, and therefore the resulting delay calculations are accurate. All other periods having actual curves that approximate the planned curve will add to the substantiation of the original schedule as a valid program.

As a final note, the most direct comparison is of actual contract work to planned contract work—without changes and change orders. You may therefore find it helpful to plot three actual curves:

1. Actual billings of contract work only
2. Actual billings of change order work
3. The combined curve of actual contract work plus changes

In a fine-tuned analysis the curves will have different uses. Even with no specialized training, however, project people will be surprised at the amount of information that will jump off the paper while they are just reviewing the plots.

## 7.6 Schedule Updating Considerations

### 7.6.1 General

The actual updating technique will vary depending upon the scheduling method selected. Bar charts only provide for the recording of the passage of time—as it passes. Other methods will allow for recording and accounting for:

- Actual activity start dates
- Actual activity completion dates (substantial completion of enough work to allow following trades to proceed)
- Time remaining to complete activities that have started but are not yet complete (as of the update issue)

Beyond this, scheduling methods and computer programs vary in their abilities and procedures for accommodating changes, adjustments in schedule logic, and so on.

### 7.6.2 Scheduling Approach

The important ideas for managers to focus on continually include the following:

1. Throughout all schedule updating efforts consider not so much "how much time has gone by" as "*when* will it be *done*."
2. Involve *all* Subcontractors and Suppliers on a continuing basis. Force them to provide complete information, and in writing if necessary.
3. Involve the Owner and design professionals. Develop a routine for transmitting schedule updates, reporting the status of the project, indicating the effects of everyone's actions and inactions, and describing "get-well" plans.
4. Use your schedule as *the* agenda at all job meetings. Let everyone see that you are *always* considering the scheduling effects of all occurrences and decisions, and displaying those effects on paper.
5. Develop a complete lack of tolerance for *any* schedule slippages. Be forceful. Don't risk the development of even the slightest bad habits on the part of any project participant.

Use the sample forms and letters that complete this section, but be creative. Use them as a guide from which to develop detailed accountability systems that are closely coordinated to the specific scheduling system used by the Company.

### 7.6.3 Sample Schedule Analysis/Evaluation Report *(page 7.26)*

The Sample Schedule Analysis/Evaluation Report that follows can be used as the routine correspondence that:

1. Transmits the schedule or schedule update to the Owner and design professionals
2. Distributes all current schedule information to each key Subcontractor
3. Reports the current status of the project as of the date of the update
4. Includes anticipated "get-well" programs intended to bring identified deficiencies back into line

In preparing the report:

1. Include names, companies, and dates. Be as specific as possible.
2. In the "Remarks" section, indicate all new constraints and conditions:

## 7.6.3
## Sample Schedule Analysis/Evaluation Report

SCHEDULE ANALYSIS / EVALUATION REPORT

Date: ______________________ Project:: ______________________ #: ________

To: ______________________ Report Date: ______________ Rev. #: ________

______________________

______________________

ATTN: ______________________

| # COPIES | REV # | REVISION DATE | DESCRIPTION / TITLE | BASELINE COMPLETION | REVISION COMPLETION | DEVIATION (+/-) | |
|---|---|---|---|---|---|---|---|
| | | | | | | | WD<br>CD |

As of ________________, 19 _____, this project is ________ Working Days (Calendar Days) ahead of (behind) schedule, primarily due to:

Remarks:

cc: Jobsite w/att.

File: Sched. Rev #_____ w/att, CF

Signed:

*a.* If the project is doing well (on or ahead of schedule), meaningful remarks might include:

- The level of confidence by management in the ability of the project to maintain such performance
- Specific critical areas that require direct attention to maintain the status
- Rough extrapolation of the performance into the future periods, if other known conditions exist whose effects have not yet been felt on the project
- Any significant facts, proceedings, etc., which might impact on the status report of the *next* or other future update(s)

*b.* If the project is in an unacceptable condition (behind schedule), include detailed outlines of:

- All confirmed commitments, promises, etc., relative to any "get-well" programs devised to this point
- Nonperformance, inattention, inappropriate action, etc., on the part of any Subcontractor, Supplier, design professional, the Owner, or whomever (be specific, name names)
- Previously "noncritical" items that have now become critical
- Specific actions required by whomever

Use names and dates. Pin it down. Refer to all appropriate correspondence, meeting minutes, and so on to support your remarks with the more detailed project record (and make subsequent research much easier).

3. Distribute copies of the completed report to:

*a.* All parties referenced in any way in the "Status" or "Remarks" sections
*b.* The Owner
*c.* The Architect
*d.* All major Subcontractors

The periodic distribution of such complete information will generally fulfill most procedural requirements that will support any requirement for timely and complete notice.

### 7.6.4 Sample Delay Letter #1 to Subcontractors *(page 7.28)*

The Sample Delay Letter #1 to Subcontractors can be used as a first formal notification to a Subcontractor that his or her work is falling off and that the schedule is accordingly in jeopardy. It:

1. Notifies the offending party that current progress is unacceptable and is adversely affecting the project
2. Indicates the specific problem activities
3. Notifies of responsibility for the resulting effects
4. Requires specific written confirmation of new commitments necessary to regain the schedule

A supply of these form letters should be on hand and completed *as each schedule update is being conducted.* When in doubt, send it out.

## 7.6.4
## Sample Delay Letter #1 to Subcontractors

**Letterhead**

To: ______________________
______________________
______________________
______________________

Attn: ______________________

Date: ______________________

Project: ______________________
______________________

Project No. ______________________

Confirmation of Fax:
Fax #: ______________________

SUBJ: Anticipated Project Delay

Mr./Ms. ______________________:

At your current rate of progress, it is apparent that the schedule completion dates for the following activities will be missed:

| Activity | Scheduled Completion |
|---|---|
| | |
| | |
| | |
| | |
| | |
| | |

If these completion dates are not regained, you will directly interfere with and delay all activities scheduled to immediately follow, thereby directly delaying key milestones and completion dates. In this event, you will be held responsible for all associated costs and other problems resulting.

Please submit your written statement to my attention by ______________, 19 ______, specifically advising of how you intent to regain the schedule. Include:

1. Number of Workers per day.
2. Number of regular and *overtime* hours per day planned.
3. Number of days for each sub-activity.
4. Specific measures being taken to expedite all necessary material deliveries.

Your complete written response is critical to the orderly completion of the project without serious incident. Your cooperation is appreciated.

Very truly yours,

COMPANY

______________________
Project Manager

cc: Jobsite
cc: File: Vendor File: ______________
Sched. Rev #__________, CF

### 7.6.5 Sample Delay Letter #2 to Subcontractors *(page 7.30)*

The Sample Delay Letter #2 to Subcontractors is to be used in those instances where a chronically deficient Subcontractor continues to be unresponsive to you and to the project requirements. It:

1. Acknowledges that no acceptable response has been received to Letter #1
2. Invokes the subcontract acceleration provisions directly
3. Notifies of the serious consequences of the continued inappropriate action
4. Again requires a written statement of planned correction
5. Notifies of at least the pending backcharge resulting from completing the work by other means

The unfortunate truth is that if this letter becomes necessary, there are sure to be other problems with that Subcontractor. The Project Manager should be at least consulted before the letter is sent out, and it is probably a better idea for the letter to go out with the Project Manager's signature. In any event, if things have deteriorated to this degree, there should undoubtedly be other communications going on to bring such a serious problem under control. If all this is happening correctly, this Letter #2 becomes an important supplement, rather than the only communication. For its best effect, it should be used as a word-for-word letter.

## 7.6.5
## Sample Delay Letter #2 to Subcontractors

**Letterhead**

(Date)

To: (Subcontractor)

Confirmation of Fax)
Fax #: ____________________

CERTIFIED MAIL
RETURN RECEIPT REQUESTED

RE: (Project)
(Company Project #)

SUBJ: Project Delay and Directed Acceleration

Mr. (Ms) :

As of this date, we have received no appropriate response to our letter of (Date) regarding your anticipated project delays. Since that date, your work has continued to progress unacceptably, causing additional interferences.

Per Paragraph ( ) of your subcontract, you are hereby directed to immediately accelerate your work by adding all labor, material, equipment, overtime, Saturday, Sunday, and Holiday work as necessary to regain the schedule.

your failure to comply with this directive will be considered a material breach of your subcontract. Appropriate action will then be taken by this office, which may include termination of your subcontract.

Please deliver your written confirmation to this office by or before (Time) on (Date) that you are either mobilizing to comply with this directive, or that you specifically refuse to comply.

No response by the date and time indicated will be construed as your notification by default of your refusal to comply. In that event, steps will be taken to expedite your work by other means. Your company will then be backcharged for all costs incurred, including but not limited to procurement and coordination time and effort, supervision, overhead, and profit. You will also be held responsible for all project delay costs and consequential effects.

Very truly yours,

COMPANY

Project Manager

cc: Jobsite
File: Vendor File: ( )
Sched. Rev. # ( ), CF

# Index

## About the Author

Andrew M. Civitello, Jr. has managed construction projects of many sizes and types for clients that include local, state, and federal government agencies and departments, municipalities, banks, service organizations, nonprofit organizations, health maintenance organizations, and private developers.

A graduate of Syracuse University, he has been a project manager for major construction companies, and president of a Connecticut general contracting firm oriented to public projects. He is an independent consultant for the disciplines of planning and scheduling, changes/claims, and project management for clients that include general contactors, subcontractors, attorneys, design professionals, and owners. He is a university instructor for project management subjects, an arbiter for the American Arbitration Association, and an expert witness for scheduling, estimating, and claims.

He is the author of *Construction Operations Manual of Policies and Procedures* (McGraw-Hill, 1994), *Contractor's Guide to Change Orders* (Prentice-Hall, 1987), *The Builder's and Contractor's Yearbook* (Prentice-Hall, 1987), *The Constriction Manager,* (Prentice-Hall, 1987 through current issue), *The Construction Operations Manual of Policies and Procedures* (Prentice-Hall, 1982), and co-author of *Construction Scheduling Simplified* (Prentice-Hall, 1985).

## DISK WARRANTY

This software is protected by both the United States copyright law and international copyright treaty provision. You must treat this software just like a book, except that you may copy it into a computer to be used and you may make archival copies of the software for the sole purpose of backing up our software and protecting your investment from loss.

By saying, "just like a book," McGraw-Hill means, for example, that this software may be used by any number of people and may be freely moved from one computer location to another, so long as there is no possibility of its being used at one location or on one computer while it is being used at another. Just as a book cannot be read by two different people in two different places at the same time, neither can the software be used by two different people in two different places at the same time (unless, of course, McGraw-Hill's copyright is being violated).

## LIMITED WARRANTY

McGraw-Hill warrants the physical diskette(s) enclosed herein to be free of defects in materials and workmanship for a period of sixty days from the purchase date. If McGraw-Hill receives written notification within the warranty period of defects in materials or workmanship, and such notification is determined by McGraw-Hill to be correct, McGraw-Hill will replace the defective diskette(s). Send request to:

Customer Service
TAB/McGraw-Hill
13311 Monterey Avenue
Blue Ridge Summit, PA 17294-0850

The entire and exclusive liability and remedy for breach of this Limited Warranty shall be limited to replacement of defective diskette(s) and shall not include or extend to any claim for or right to cover any other damages, including but not limited to loss of profit, data, or use of the software, or special, incidental, or consequential damages or other similar claims, even if McGraw-Hill has been specifically advised to the possibility of such damages. In no event will McGraw-Hill's liability for any damages to you or any other person ever exceed the lower of suggested list price or actual price paid for the license to use the software, regardless of any form of the claim.

McGRAW-HILL, INC. SPECIFICALLY DISCLAIMS ALL OTHER WARRANTY, EXPRESS OR IMPLIED, INCLUDING BUT NOT LIMITED TO, ANY IMPLIED WARRANTY OR MERCHANTABILITY OR FITNESS FOR A PARTICULAR PURPOSE. Specifically, McGraw-Hill makes no representation or warranty that the software is fit for any particular purpose and any implied warranty of merchantability is limited to the sixty-day duration of the Limited Warranty covering the physical diskette(s) only (and not the software) and is otherwise expressly and specifically disclaimed.

This limited Warranty gives you specific legal rights; you may have others which may vary from state to state. Some states do not allow the exclusion of incidental or consequential damages, or the limitation on how long an implied warranty lasts, so some of the above may not apply to you.